Inga's Amazing Ideas
by
Ann Rubino

Inga's Amazing Ideas

All rights reserved. Printed in the United States of America.

No part of this publication may be used or reproduced in any manner whatsoever -- mechanically, electronically, or by any other means including photocopying-- without written permission, except in the case of brief quotations embodied in critical articles and reviews. For information address all questions to the publisher at catree.com.

Text and sketches copyright © 2018 Ann Rubino

Inga's Amazing Ideas/Ann Rubino

978-1-942247-10-4 (paperback)
978-1-942247-11-1 (hardcover)
978-1-942247-12-8 (digital)

1. Fiction : Historical - General 2. Juvenile Fiction : Historical - United States - 19th Century 3. Juvenile Fiction : Science & Technology

1. Rubino, Ann.

10 9 8 7 6 5 4 3 2 1   Published by www. catree.com in USA

Dedication

For Amelie, who wanted a story about a girl, and in memory of Anna Alexandersdotter Johansson, my "forsmor", who came from Sweden long ago.

Books in the <u>Floyd County Chronicles</u> Series:
Emmet's Storm
Inga's Amazing Ideas

Other Books By Ann Rubino:
Peppino, Good as Bread
Peppino and the Streets of Gold

Contents

Chapter 1
Floyd County, Iowa

I won't cry, Inga muttered to herself. Papa had said, "Be brave, head up, like the Vikings," when he signed her over to Mrs. Hobson of the Children's Aid in New York. Papa thought of Vikings, but Sweden and mother were distant memories. There's nobody here worth crying about, she thought. Dusty little town. Old wooden train platform. Not even a decent sidewalk. Bunch of sunburned farmers and their straggly wives walking up and down along the train platform to choose a kid to live with them. They always like the pretty little ones with hair bows. The strong boys were all taken by the time we passed Chicago. It's not such a great thing to be wanted by somebody chewing on straw.

Smoothing her faded blue gingham dress, Inga straightened her shoulders, and placed her worn carpetbag neatly next to her right foot. She stared with her chin up, across the track into the distance.

"Pretty slim pickin's, Lavinia," Henry Duffy remarked to his wife as they walked along the line of children. "Hiram said the train would bring a lot of good strong boys from the cities looking for a place." They walked along the track, past several toddlers who looked scared and a thin boy of five scuffing his shoes in the dust. "All I see is a half dozen leaky-nosed tots. There's just one boy and he can't be more than five years old. Just him and this skinny girl," Mr. Duffy said. They stopped near a tall, thin girl. She had a stern look on her face and hair so light it was almost white. He gestured with his head at Inga.

"A girl might do," Mrs. Duffy said. "She could help me too."

"Help you? I'm the one who needs the help," her husband Henry, the blacksmith, retorted. "A boy could work the bellows to keep up my fire, and hold the horses when I nail on their shoes. You just have the house and the pies."

"We may have to wait for another train then, Henry."

Mr. Duffy chewed on his pipe stem and sighed. Silence.

"All right, Mother." He looked up and down the row of children.

"What about this one?" he asked, turning back and pointing with his pipe toward Inga. "She's old enough to help you with your pies, take some of your load off. Not much use to me though. I'll just have to suffer with my lumbago through another season." He reached out to feel Inga's arm for muscle. "Skinny. What do you think, Mother?"

"She's a child, not a horse. I'm sure she can speak for herself." Lavinia Duffy looked down at Inga, who was blinking back big tears in her light blue eyes. "What can you do, dear?"

"Clean, cook, carry water, wash, sweep." Inga stared straight ahead as she said this.

Henry Duffy looked down at Inga's carpetbag. It was old, but the handle was repaired neatly with wire woven in a herringbone pattern. It was a clever repair, almost like a decoration. "Who fixed your carpetbag, girl?"

"I did."

The train gave two toots of the horn. "Form up your line, children," Mrs. Hobson called. "We have yet another stop today. Oldest go on ahead to help the little ones. Coming, Inga?"

The children turned and followed Mrs. Hobson toward the train steps, Inga guiding the little ones up the steps. "There goes our chance for a worker," Mrs. Duffy told her husband. Mr. Duffy took his pipe out of his mouth and waved it.

"Wait a minute, girl," he called, lumbering after her. "What else can you do?"

"Fix things."

"Fix what?"

"Fix anything. Papa taught me." She didn't look at him.

"Hold the train," Henry Duffy called. "We'll take this one." He hurried over to Mrs. Hobson. "Where do I sign?"

Mrs. Hobson bent down and pulled out a sheaf of papers from her battered brown carpetbag. She paged through them. "Flournoy, Graham, Hagstrom,

Keely, Patterson—aha, that's it--Hagstrom, Inga. Here it is." She put that sheet on top of the bundle and held it for Duffy. "Sign here," she said. "You're responsible for her food, clothing, shelter. Schooling until the age of twelve. Only one more year in her case. Once she is eighteen, she's free to make her own life. She's to write a report to me once a year at the address on top. If she has relatives, I'll pass it along to them. If it is reported that you mistreat her, our representative will remove her from your care and find her another place."

Henry Duffy signed in a scribble. "Come along now, Inga," he said, heading back toward the end of the station.

Mrs. Hobson patted the child on the shoulder. "Take your carpetbag, now. Be a good girl for your sponsors. And good luck to you." She turned and joined the children on the train.

Inga picked up her little carpetbag and followed Mr. and Mrs. Duffy as they headed away from the train station toward the small white house that shared a wall with the blacksmith shop. They passed two boys sitting at the corner of the platform. "Hello, Emmet, Tom," Mrs. Duffy said as they passed. "What mischief are you two up to today?" The boys smiled sheepishly.

"Looks like slaves," Emmet muttered to Tom as the Duffy group went out of earshot.

"Naw. They don't have slaves any more."

"I know that. President Lincoln gave an order that they had to let them go. That was only for the states in secession. Then in 1868 Congress passed an amendment stating..."

"Emmet, for grief's sake leave off. All your knowledge gives me the whim-whams. You know they ain't slaves."

"Well, it looks like. All lined up for people to pick."

Walking along the road, Mrs. Duffy slowed to match Inga's pace. "You will be safe with us here. Mr. Duffy is grumpy but he will not beat you."

But he's not my papa, Inga thought. "What means 'grumpy' "? she asked. "It is bad?"

"His voice," replied Mrs. Duffy. "He shouts. Sometimes he pounds on the table." She paused and looked down at Inga, trying to see the situation through a child's eyes. "I will take care of you," she said. "You will have a good home here."

They walked on for a while, avoiding the mud puddles on the road. Inga noticed how neatly Mrs. Duffy lifted her skirts to avoid the horse droppings. In New York sweepers cleaned the streets every night. The smell of beer and sweaty leather wafted out of the saloon as the swinging doors closed behind a farmer.

They passed a house with wires coming out of the roof. A sign said: Post Office, Floyd, Iowa. Below was the inscription: H. Roche, Telegrapher.

Kelly's General Store had a rail across the front for hitching horses and two little boys were trying to climb up to balance on it. At the corner another dirt street branched off to the left, away from the railroad tracks. A church stood on that corner with a small school building in back. Low wooden houses lined the street, some painted and some plain gray wood. One had a gold-lettered sign on the front, "John Egan, M.D." A picture of a skeleton was mounted on an easel inside the window along with several blue and brown bottles marked Rx.

This is so different, Inga thought. No streetcars,

no bricks on the streets. Only the saloon smelled like the city. "Do you have a house?" Inga finally asked. "In New York we lived up five sets of stairs. These houses look short."

"You're out of the city now, Inga. All the houses are short, as you say. Our house is up ahead of us." She pointed up ahead. "There, do you see it? It's the white one with a yellow door. We have a loft, though, where you will sleep. This is the whole town. There is no more."

"What is loft?"

"It is a place on top, over the other room. There is a ladder to climb. You will have your own bed and it is quiet. It will be your own place."

Mr. Duffy turned around and saw them lagging behind. "Get a move on, woman. The day is going fast. Work won't do itself," he grumbled.

"Is that grumpy?" Inga asked.

Mrs. Duffy laughed and called back to him. "I'm well aware of that sad fact, husband. I've yet to see a pie that set itself in the oven, or a dinner that cooks itself. I just want Inga—is that your full name? I never heard it before -- to feel at home."

"Ingeborg Hagstrom. People call me Inga."

"She'll understand soon enough," Duffy continued, turning his head back toward them. "It's no great mystery. Just come along."

Once past the church, they saw the Duffy house at the end of the street. Its paint was peeling in some spots, but the yellow door gave a cheery look. The two rosebushes near the front door had lost most of their leaves this late in the summer. There was a window next to the door and another one up higher, close to the slightly sagging roof.

A big sliding door stood open to the rough-looking shop. Inside, tools lay scattered on the heavy work table and the dirt floor. Other than the sun shining through the small window and the dark red glow from a charcoal fire, the shop was dark. A man stood inside, loosely holding the bridle of his horse.

"I thought you'd moved to China," Jim Ronan called out over his horse's back. "Belle here has been waiting for a new shoe this past hour."

"We were at the train."

"It seems you have a young visitor. Is she your kin?"

"Nope. She came off the Orphan Train from New York, needing a home and work. She'll be with us and help out the missus until she's eighteen."

"I would have thought Mrs. Duffy was at least twenty, Henry. Though she's pretty as she was at eighteen," Jim teased.

"Aw, Jim," Mrs. Duffy said. "You're always full of the blarney. You know I went to school with your mother. Henry is talking about the girl. Her name is Inga Hagstrom and she's placed with us. Say hello, Inga."

"Hello." Inga peered around Mrs. Duffy to get a better look. She looked past the massive horse and the tall farmer with his mop of blazing red hair, and noticed the tools. What a wonderful place she thought: every kind of tool, and a big place to use them. What grand things could be made in this shop.

"Come inside and see my shop, Inga," Mr. Duffy called. "Horses need shoes. You know that. I make them and nail them on. Come in and I'll show you."

Jim held out a calloused hand to lead her into the blacksmith shop. He handed Inga's bag to Mrs. Duffy standing on the road outside.

"We'll be in to supper shortly," Mr. Duffy called back to his wife. "I'll just take care of Ronan's horse here."

"No hurry, then, Henry. That supper may take some time to make itself." She trudged inside carrying the carpetbag. "Help, my foot," she mumbled. "Another mouth to feed and more washing. I wonder what I've let myself in for. Yet she's a quiet thing and I think she may be intelligent. She noticed the houses in town right off and started trying to understand life here. It's a shame the little lass had to leave her home and come all this way. What could make parents let go of her?"

After seeing Duffy's shop, Inga went into the house. She was surprised to see it was mainly one very big kitchen. Along the back wall stood a big black iron stove. The stove had two ovens, each with its own fuel door. Pieces of wood were stacked neatly at each end ready to feed the fires. Two pots of hot water for washing sat on top. A long heavy work table with pie pans stacked at one end ran down the center of the room. There was a bed in the corner, covered with a bright patchwork quilt. It was separated from the rest of the room only by a thin rose gingham curtain. A wooden chest stood alongside, holding a kerosene lamp and a water pitcher.

Inga's elbow brushed a half-empty sack of flour on the table. "Careful there," Mrs. Duffy warned. "That will be going into food." Cooking tools lay scattered on the surface. Four fresh pies were lined up neatly in a row at one end.

The sink had a hand pump for water, and a pipe draining down through the floor. Work clothes hung on pegs in the walls and shelves held jars of fruit and other food supplies.

"You cook very much," Inga said. "Your man, he does eat many pies?"

"Call him Mr. Duffy, and he does, but these pies aren't for him. I sell pies, some of the best in town."

Inga looked at the bed again. "Where is my bed?" she asked.

"Oh, goodness me. Let me show you." Mrs. Duffy walked over to pull a ladder out from back of the stove. She lined it up with the square opening in the ceiling. "Here's your place," she smiled, wedging the ladder top into notches in the ceiling. "All to yourself. We keep sacks of flour up there too, but they're no bother. The mice don't get at it as much upstairs. You can cuddle up in your blanket on the straw mattress. The pegs in the wall are for clothes and I'll find you a water pitcher. It will be lovely." She turned back toward the shelves.

"Mice?" asked Inga.

"Just a few. They're very quiet. I just can't keep them away from the flour, you know. I suppose they have a right to live."

"Do they have to live in my room?"

"My, you're pretty persnickety for a person who's just been offered a warm bed and a roof overhead."

Inga labored up the ladder dragging her carpetbag. I guess this is my life, she thought as she plopped down, tired, raising a cloud of straw dust. Two sheets and a blanket lay folded down at the end of the mattress behind some flour bags. A small pillow, tin basin and towel were with them. And sure enough, there were several pegs in the wall for hanging clothes. Maybe she could find a chair someplace—maybe trade for it—and someday a table. A kerosene lamp too, so I can read, she hoped. I'll read my father's letters. I'll write and tell him everything that happens.

"Anything to eat yet?" a voice hollered below. Mr. Duffy was home and hungry.

"Yes, husband, I'm making johnnycakes," Inga heard Mrs. Duffy reply. "I make them for supper every Saturday," she called up through the opening. "Mr. Duffy is partial to them. Just rest yourself a bit. There's cool water down here in the pitcher when you want it. I just pumped it fresh."

Little cooking sounds drifted up the ladder— the tap and scrape of a spoon beating eggs and flour, the fresh sizzle of batter hitting a hot pan, then the wonderful smell of fresh johnnycakes and – could it be? -- apple pie. "Come to supper," Inga heard. "Food's ready." It wasn't a moment too soon for Inga. She had shared her last sandwich on the train with a little boy from the Bronx, so her last meal was yesterday morning. Pie might fill the lonely feeling inside.

Chapter 2
Sunday

"We must go to church early," Mrs. Duffy said as she hurried Inga to eat her boiled eggs the next morning. "We'll find the nun and sign you up for school. They've already been studying for three weeks. Wear your best dress. Everyone will be looking at a newcomer."

"I have this dress only," replied Inga. "And my jacket. Only small case I can bring on train."

"Oh, my," Mrs. Duffy sighed. I feared as much. She comes to me almost as bare as a babe, and with fall coming on. She patted Inga on the shoulder. "Let's have a look at you." Inga stood up, smoothing her dress and patting her braided hair. "Now turn around." She did. The back was no better than the front—the handed-down dress had seen a hundred washings, and was made for a taller, plumper girl. Her neatly-mended white apron hung at a slant, weighed down by objects in the pocket. At least her collar was clean.

"You have pretty hair anyway, if we could curl it a bit. That braid around the top of your head looks like a straw crown. It needs a bit of ribbon..." Mrs. Duffy turned Inga around again. She went to the chest and dug down into it, dislodging a couple of her old dresses and a pair of long johns. "Mr. Duffy will be livid if we're late." She found no ribbon.

"My hair is good," Inga explained. "I washed it in New York and I brush it every day. It is Swedish hair. It does not curl."

"I fear you are right," Mrs. Duffy agreed and sighed, "Just another cross to bear."

"Not cross, a braid."

"Oh, never mind, we'll be late. Just take the things out of the pocket that pull your apron askew. We have to go."

"I must not lose my tools." Inga pulled out a folding knife, screwdriver and pliers.

"Oh goodness. You're as bad as Mr. Duffy! Just leave them here on the table."

Mrs. Duffy grabbed Inga's jacket and stuffed her into it, licked her finger to smooth out Inga's hair crown, and headed for the door.

Mr. Duffy was pacing out in front of the house, kicking his boots through the leaves as he went. "Get a move on there, woman! If you plan to sign this child up for school we need time to talk to Sister Ludwig. She'll walk away from us as soon as the church bell rings."

Mrs. Duffy grabbed Inga by the hand and headed onto the street. "We're off then. Lead on, husband. I'm sorry to hold you late."

In front of St. Patrick's, Sister Ludwig stood talking to some of her students and waved to the Duffys as she saw them approach. "Good morning. Whom do we have here? A niece perhaps?"

"Just an orphan," Duffy said. "We got her yesterday. She's to help my wife." He spotted Hiram Roche and walked over to tell him, leaving the women to talk.

"I want her to go to school," his wife answered Sister. "She's not just a helper."

"What is your name, dear?" Sister asked. Inga was silent. "Can she speak? Is she a deaf-mute?"

"She's Swedish and very shy," said Mrs. Duffy. She turned to Inga. "Tell Sister your name, dear."

Inga pulled Mrs. Duffy's head lower by her bonnet strings, to talk into her ear. "Why does she wear that hat?"

"The nuns wear them. It's like a uniform."

"Her dress looks funny."

"It is the dress all the nuns use."

"She is your sister?"

"And no, she's not my sister. We call her that out of respect. She is nice. You can talk to her." Mrs. Duffy straightened up, adjusting her hat. "Ask her again, Sister. If you talk slowly she'll understand."

Sister smiled slightly and asked again. Inga answered.

"My name is Inga Hagstrom. I live with Duffy family. I help them. I don't bounce."

"She doesn't bounce?" Sister asked Mrs. Duffy. "How odd. Have I misunderstood?"

"Oh no, Sister. Someone told her people were looking for bouncing babies. Perhaps she thinks bouncing is a requirement. She is only in America less than three years, and New York at that. You know how badly they speak there. I know she'll learn plenty of good American English in your school. I want to enroll her."

"You're making a real investment in a stranger's child, Mrs. Duffy. Does your husband approve?"

"Of course not. You know he would squeeze a nickel until the buffalo on it bleeds. I'll pay with my pie money. A child needs an education, whether she's mine or no."

"If she can read, I can put her in the middle class."

"She reads and writes."

"Swedish," Inga said softly. "I learn some English in New York. I can read Swedish Bible."

"Well, then..." Sister seemed lost in thought for a moment.

In the distance, Mr. Duffy, Emmet and his father Hiram, the postmaster, spotted Sister and headed towards them. Duffy's voice carried down the street. "We got this girl, Hiram. They were all out of boys."

"Oh, dear. Here comes Emmet Roche. He and Tom Kelly are always looking for things to meddle with. You know them, Mrs. Duffy."

"I remember them well. They lit Brady's hay wagon on fire with their reckless hot air balloon demonstration."

"Just bring the girl over tomorrow at 8:30 and we'll see what can be done."

The church bell rang just as a gaggle of young girls rounded the corner chattering and swinging their skirts. "Look what the cat dragged in," one said rolling her eyes at Inga and thinking no one heard. The others giggled.

"For shame, girls," Sister said sharply. "You have the nerve to ask the good Lord for His help after you shame one of his poor creatures! I'll be speaking with your teacher tomorrow. Perhaps you need to write a hundred lines about this in your journals. Report to my classroom tomorrow as soon as the bell rings."

They all went into the church, Sister Ludwig pursing her lips severely, the girls sullenly mumbling to themselves, Inga clinging to Mrs. Duffy's hand. They don't like me already, she thought sadly.

Dearest Father,

I write to tell you of my journey. I was so frightened
at first of Mrs. Hobson and the train with its frightful
noise. But I did not cry or scream like the little ones.
They were so afraid. Mrs. Hobson let me help her and
the nurse. I was like a big sister for them.
Every day on the train we went for many miles. It was
very warm and stuffy, but open windows were
forbidden because smoke and dirt fly in and harm the
little ones. At every station we had soup from a big
kettle. If the town had farms we had fresh milk.
At each stop, people came to choose children. Some
had letters with the child's name already written. Some
had no papers and Mrs. Hobson helped them pick a
child they wanted. Most wanted strong boys. They also
wanted pretty little girls with curling hair. The big
boys were all gone by the time we got to Chicago.
At Floyd, Mr. Duffy and Mrs. Duffy wanted me. They
liked my carpetbag! I don't understand how they think.
I work very hard here and try to be brave as you told
me. I miss you very much. I don't miss living in the
rooming house. I have a bed to myself in a quiet loft
that is all my own.
Please write to me and say you didn't forget me.

Your Dotter,
Inga

Chapter 3

Making Pies

"I'm going out to Brady's farm, Lavinia," Duffy called from the shop doorway. "He needs some help hoisting the hay bundles up into the barn. Too bad his boy Cyril is useless now, without his foot. It makes it hard for Brady." He climbed up into his wagon, snapped the horse's reins and set off down the street.

"What means 'useless'?" Inga asked Mrs. Duffy.

" 'Useless' means can't do any work. He had to have his foot cut off."

"Cut off? No foot at all?"

"No. Not a bit. Here, help me shift this sack of flour closer to the table. Can't waste our day talking about that good-for-nothing."

"What...?" How can they cut someone's foot off, Inga wondered, but Mrs. Duffy was in motion and couldn't be stopped.

"Get a move on here. We have work to do. Ten pies ordered today. You can peel the apples." She handed Inga a short knife and pointed to a bushel of apples. "Just peel them, cut them and pile the pieces in this tub."

It was quiet for a while. Mrs. Duffy put a huge bowl on her work table and poured in a peck of flour. She weighed out two and a half pounds of lard from a big brown crock and threw it in with a handful of salt. Then, with a knife in each hand, she chopped the lard and flour together until it was all crumbs.

Inga peeled apples, one after another, dropping peelings on the floor next to her, and after an hour the tub was filled with apple pieces. She rubbed her wrinkled and aching hands as she said, "I finish. I am happy to cut all apples."

"All the apples, dear? No those are just today's. The shed out back is full of apples. Did you think I have to depend on one little tree?"

Mrs. Duffy pointed with her chin towards the pump at the sink. "Get me some water now. I'll need it to finish the dough. Pour some for yourself as well." Inga pumped two glasses and brought one to the table. "Just dribble it over this flour."

"What means 'dribble'?"

"Like this..." Mrs. Duffy sprinkled the whole glass over the flour. "Now get me another."

She only sprinkled part of this glass and started mixing the dough with her hands. Pretty soon it got sticky and firm. She dusted some flour on her hands and straightened up, rubbing her back. "Lucky it's cool today," she remarked. "When the weather is hot the dough is hard to handle. Now get me that big rolling pin from the sideboard and we'll set to work."

"It is not done?"

"Oh no, my lassie. Far from it. Now we have the heavy work. Just watch." She took a lump the size of a small cabbage out of the bowl and smoothed it. She sprinkled flour on her table and smacked the lump down. She held out her hand and Inga gave her a rolling pin. Then back and forth she rolled it over the dough, turning it and dusting it with flour every couple of times. It flattened out to a big circle.

"Now get one of those pans." Mrs. Duffy pointed to a stack in the corner.

She draped the dough over the pan and put the pan at the end of the table. "Now the next. Only nine to go."

Inga looked at the tub of apple pieces and the loads of dough and blinked. "There has to be a better way to do this," she said. "It will take all day."

"Of course. It's a day's work. Did you think we brought you here for a rest? Here, stand at that end of the table and do what I do. You can make the tops." She handed Inga another rolling pin and shoved a handful of flour to the empty end of the table. They both set to work. "You want to have a quick light touch," Mrs. Duffy advised. "Just a light roll or two, then move it gently but fast, like this, then another couple of rolls. Try to get the dough to stretch out every time."

After an hour ten dough-lined pans stood in a row, waiting for apples. In went the slices. Mrs. Duffy sprinkled the pies with sugar and slapped a pat of butter on top. Inga covered them with dough. Mrs. Duffy crimped the edges and poked a pattern of holes on the tops. " 'A' is for apple," she explained, "so I remember what's inside. "I put 'C' for cherry, 'M' for mince and so on. You'll learn that quickly."

By this time the fire in the stove was getting low. Mrs. Duffy stoked it with more pieces of wood. She loaded five pies into the oven and looked at the clock. "One o'clock, Inga. In half an hour we'll move them around. By two or so, we should have the first batch finished. We can deliver them while the second batch cooks."

"Do we eat?" Inga asked.

"Oh, I forgot. With Mr. Duffy gone I plain forgot to feed you. Just have a piece of bread. There's butter and maybe a bit of ham in the keeper. You may eat one of the apples if there are any left."

"You do this every day?"

"Almost every day. Tomorrow I'll wash. Then the next day I'll make more pies. It's hard to keep up with the baking and all. I'm glad to have your help."

"When I will go to school to learn the English?"

"You can go when I don't need your help. We'll try for three days a week. Education doesn't do much for a girl once she's past the reading and writing."

The door flew open and Mr. Duffy marched in. "Still making them darned pies, I see. Lettin' all your work go. I've been waiting a week for a patch on my pants and the washing is piling up. We couldn't finish stowing Brady's hay. I'll have to waste another day. His boy Cyril just limps around. All he can do is hold the pulley rope while we wrestle the bales off the wagon. I feel bad for Brady. Is there any speck of food to be found?"

"I'm sorry, Henry. I got off to a slow start. I'll rustle you up a lunch right away."

"Blamed woman. Always running behind." He pulled a stool up to the table and plopped down on it, getting his elbows full of flour. He brushed them off. "Can't even keep the kitchen clean."

"I sell them for 12 cents each, Henry. That's real money."

"Egg money's more like it. Don't know where you'd be if I didn't break my back every day to keep you in fripperies."

"Excuse please. What means fripperies?" Inga asked.

"Women's gewgaws. Ribbons and frocks and hats and such. Plain useless stuff that I sweat for."

Inga looked around and wondered. Where does she keep the fripperies? Mr. Duffy addressed himself to his plate of eggs and ham and didn't say any more.

Chapter 4

First Day at School

The whole upper form class at St. Patrick's swiveled their heads as Mrs. Duffy arrived at the door with Inga. Inga wore her same gingham dress that Mrs. Duffy had smoothed with an iron the night before. Her white-blond hair was neatly braided into a crown around her head and her old shoes neatly blackened with stove polish. Her crisp, clean apron sagged at the pocket. "This is Inga Hagstrom," Mrs. Duffy said. "I can spare her a for few weeks before the holiday rush."

"Good morning, Mrs. Duffy," said Sister Norah James, the teacher of the older children. "Before Halloween, Sister Ludwig told me you would be coming in soon. We have gone on ahead with our work now. Of course I'll accept her, but it may be too difficult. Sister Ludwig told me she only reads Swedish."

"She's had a lot of work to do," Mrs. Duffy replied. "No time for reading. You know she's an orphan and needs to earn her keep. She's intelligent and obedient, though. I think she'll be able to learn. If she cannot, you can move her to the lower form with the little ones."

Out of the side of her eye Inga spied Emmet in the front row, next to the teacher's desk. He stayed still but waggled his fingers in welcome.

Sister Norah James looked down at Inga. "Do you know any of these children, Inga? What a strange name. I never heard of a Saint Inga. Are any faces here familiar?"

"I do not play any children."

"I see. Your English is better than I had feared. I'll seat you next to Emmet and he can help you with your words. He has a superabundance of them." The class tittered and Emmet's face turned pink. "I will keep you abreast of her progress, Mrs. Duffy," Sister Norah James said. "I commend you for your charity in offering her an education."

Mrs. Duffy said goodbye and went to complete her errands. Sister moved another student back to an empty seat and pointed Inga to the place next to Emmet. She tapped on her desk with her ruler and touched the writing on the chalkboard that said "Reading 9:00-9:45." The book monitor gave out the readers, which covered three grade levels. Book 6 went to Emmet who could read anything that they could print. Next to Inga, the monitor looked up at Sister, inquiringly. "Fourth," she said quietly. "Just to try." He handed Inga McGuffey's 4th. "Read quietly to yourselves, children," Sister said as she stood behind Emmet and Inga. "Emmet, read the first page of Inga's book very softly to her so she can hear what the words sound like. Then she will read each sentence back to you. If she has a problem with the word, explain it to her. I must tend to the rest of the students." She turned back to the rest of the class.

A soft murmur of reading floated through the room as the students worked through their books. Emmet's seat squeaked every time he shifted around, and Emmet was a shifter. Twice Sister had to give him her "teacher look" and once smacked her ruler on the desk nearest her, scaring Ozzie Crowder out of his wits. Softly Inga stood up and tapped Emmet, motioning him to stand up as well. With a puzzled look he stood up. Inga slid her screwdriver out of her apron pocket, knelt down and tightened the screws under the seat. Before Sister reached the front of the room they were back to reading, with Inga's screwdriver safe in her pocket. "Why were you out of your seat, Emmet?" Sister whispered severely. "You're supposed to be reading."

"I fix chair," said Inga. "Please excuse."

Emmet wiggled vigorously, almost knocking the books off the top. "The squeak is gone, S'ter."

"That will do, Emmet. Let us try to do repairs during recess, however." She put her hand on Emmet's desktop and wiggled it hard. There was no squeak. "Children these days... not like they used to be," she muttered as she walked to the back of the class shaking her head.

Emmet helped Inga every day that she came to class, usually on Mondays and Wednesdays. Other days were taken up by pie-making and other work, but little by little Inga found she could read more English. "Is this all we do?" she asked Emmet one day. "Only words?"

"School is all about words," he replied. "Did you think it would be fun? It's words and more words. But they're good when you find something interesting like *Scientific American* or encyclopedias. That's when they come in handy. I don't suppose they're much use for pies, though."

"We can write down how to make them."

"I suppose so. But who would want to read that? Women already know and men don't care."

Chapter 5
The Anvil

"I have to go back to Brady's again," Mr. Duffy grumbled two days later. "Those hay bales can't be left out to weather. Don't know why in tarnation he can't get those girls of his to give him a hand."

"They're little, Henry. Hannah can't be more than eight years old, and a girl on top of it. Don't grumble so. There's no smithing lined up for today anyway."

"Right. And no money coming in either." He grabbed his jacket off the hook, jammed his hat on and went through the door to the shop, slamming the door behind him. "What the..." was followed by a loud crash. There was sound of a scuffle, then silence. Mrs. Duffy and Inga heard him saddle up his horse, and then listened to the clatter of hooves echo down the street.

No more than two minutes later there was

a knock at the door. Tom and Emmet stood on the porch with their arms full of empty pie pans. "We're bringing these back," Tom said. "Mother wants two more for tomorrow. What's wrong with Mr. Duffy? He rode away like the Rebs was after him. Almost ran us down."

"He's off to help Mr. Brady hoist his hay into the barn loft." Mrs. Duffy turned to Inga. "Keep on peeling dear. We have two more orders. Come in, boys. The wind is mighty fresh outside."

"Hello, Mrs. Duffy," said Emmet.

"Feels good to be in here smelling the hot pies. It's getting frosty outside," said Tom.

"I think Mr. Duffy break something in the shop," said Inga, looking up from her tub of apples.

"Whatever it is, these pies can't wait," said Mrs. Duffy. "I'll have to see to it when I finish rolling out this batch."

"Let's go see," Tom said, rubbing his hands together to warm them. "Can we, Mrs. Duffy? We won't mess anything up in there. Just go and report."

"I don't see any harm. Go with them, Inga. The apples will wait for you. Find out what happened."

In the shop, Duffy's anvil had fallen off its base and lay sideways on the dirt floor. A long strip of dark gray cloth hung from the pointed tip. Inga picked it up and put it in her apron pocket.

"It looks like he caught his jacket and knocked it down," Emmet said.

"Any *eejit* can see that," Tom retorted. "The issue is what can we do about it. That anvil must weigh a couple hundred pounds. He'll need help to put it up when he comes back."

"Maybe we can roll it up a ramp," Emmet suggested. "I'll get a couple of rolling pins from Mrs. Duffy and find a board. We can make a ramp like the Egyptians." He hurried toward the house door.

Tom wrapped his arms around the anvil and tried to shift it. It wouldn't budge. "I think we need more help. It may take a couple of men."

"We can..." Inga said. "Tom, see!" She pointed to a large block and tackle hanging on a nail in the wall.

"What's that?"

"*Blockera och tackla.* It will make easy. I can show you. They have on ship. There is trick," Inga said. "I show you. Is rope here?"

Tom scrounged around and found a long roll. "Here. What're you going to do?"

"Watch." She wound the rope around and over, making little crossings and tightening here and there. Soon it looked like a rope net was around the anvil. She tied the last knot and jerked it tight. "Now get..." She pointed. "I have not American name. They have on ship. Pull up sails. Do what I tell you. You will find surprise."

Tom climbed up on top of a table and brought down the set of pulleys. There was an iron circle at one end and a hook at the other end of the set, with pulleys and ropes in between.

"We must hang from strong place. Something must carry." They looked around and, sure enough, there was a strong iron hook screwed into the ceiling beam. Somebody must have had this idea before. "You go up, Tom?"

Tom was willing. He dragged a stool over next

to the anvil base and climbed up. Stretch as he might, he couldn't reach the hook. Inga found a wooden crate and they balanced it on the stool. It wasn't tall enough. "Put it up on end," Tom suggested. Up he climbed, teetering slightly. After three tries he finally got the ring on the top pulley over the hook in the beam.

"Now bring down all the pulleys slowly," Inga said. "Keep them together with the rope. Don't let the rope slip off the sides."

Bit by bit Tom worked the rope up and around each pulley. "There's a mile of rope here," he called down after about a yard of rope only moved the pulley down and inch or so.

"Just keep going. The rope will do the work."

Finally the block and tackle was down near the anvil. Inga fastened the bottom hook through the rope net she had made. Tom climbed down. "Now we lift. Maybe one hand."

"You're joshing."

"Stand here." She stood him next to the base. "Pull and pull. Don't let go!"

Tom started pulling. He pulled and pulled, yard after yard of rope. "Nothing is happening," he complained.

"Yes. Is getting straight. Soon it will move. Keep going. Are you tired?"

"Naw. It's easy. It's too easy. It isn't going to work."

Emmet came through the door just then with a rolling pin in each hand, to see the anvil begin to sway gently in the air. Mrs. Duffy, behind him, covered her mouth quickly to stifle a gasp.

Little by little the anvil was lifted up and when

it got level with the base, Inga gently pushed it over into place and unhooked the device from the rope net. "Now you stop. Pull all rope up. Keep pulleys together." Tom climbed atop the teetering boxes again to take off the hook. They rolled all the rope up and put the block and tackle back on its hook on the wall.

"Duffy should be happy," Tom said. "We took a load off him."

"Without a ramp!" Emmet said. "I've read about that but never tried one. Smart idea, Tom!"

"Inga figured it out. I did the dangerous part," Tom admitted. "It was a team."

The anvil stood in its usual place, just a little crooked. Mrs. Duffy invited them in for a snack.

"Close the shop door behind us, children," Mrs. Duffy said. "That draft is fierce. I hope you returned Mr. Duffy's things to their proper places. He's fussy about his tools being touched."

"He won't even notice," said Tom.

"Here, have a bit of pie," Mrs. Duffy said. "I'm making mince meat today for Thanksgiving coming, and I made a little pie for ourselves out of the scraps to see if it's sweet enough."

They had no sooner taken a bite of their pie when they heard Mr. Duffy come into the shop, stop for a moment and then storm into the kitchen. "Who's been meddling in my shop?" he demanded. Tom and Emmet pointed at Inga. "Can't be. Somebody picked up a 200-pound anvil and put it up on the base. Inga can hardly lift a rolling pin. I want to know who came into my shop."

"No one came here, Henry. I sent the children to see what had fallen. Now we're eating pie. Would you like some?"

"Stop changing the subject, Lavinia! I want some answers. I can't have strangers mixing about in my private business. There are expensive tools and all, just there for the taking. A body could lose his whole livelihood while you're all in here eating pie."

"We help you," Inga said softly. "We make you pleased. We didn't take anything. Nothing broken. We put all ropes away."

"All the ropes? What ropes?"

"Inga showed me how to work the pulleys," Tom burst in. "It took miles of rope."

"We planned a ramp at first," Emmet corrected. "Then Inga spotted the pulleys."

"*Blockera och tackla,*" Inga offered.

"Block and tackle. Right. That's what I used when I set up my shop," Mr. Duffy said. "Well, I'll be darned, a girl engineer. Who would have thought? Of course it isn't perfectly straight. I'd say it's about and inch and a half out of square. But I can lever it over that bit. But just stay out of my shop now, all of you. You could have been hurt swinging that weight."

"We didn't swing it. It just came straight up," Tom said.

"And I push in place," said Inga.

"A bit of pie, then, husband?" said Mrs. Duffy. "No harm done. The children won't play there any more."

"Not play, Mrs. Duffy," Inga said. "Only help. I find this." She took the strip from the jacket out of her pocket and held it out. "You lose, Mr. Duffy?"

"Humph," said Duffy. "Looks like my jacket. Can you fix that too? You may work out here after all." He took a big bite of warm mince pie and stretched out his feet toward the warm stove. Watching Inga line up her apples to peel them faster, he just shook his head slowly.

Later that night, once the pie making was done and dinner was over, Mrs. Duffy heaved a sigh of relief. "Off with you now, Inga. You've had a hard day." She took a hot brick off the stove top and wrapped in a towel. "Here, put this in your bed. It gets chilly up in the loft."

Inga took the brick, nodded good-night to Mr. Duffy, and scrambled up the narrow ladder to the loft. He was lounging in the rocking chair smoking his pipe, his feet propped on the fender of the stove. Mrs. Duffy pumped water over her chapped red hands and looked at the pie pans and dinner dishes stacked in the sink, on the counter, on the table, all needing a good scrub. She swung around for a kettle of hot water from the stovetop and bumped Mr. Duffy's foot, which was warming at the oven door. "Watch your step, woman," he grumbled, and took a deep drag on his pipe. He shook open his newspaper and went back to rocking his chair.

Mrs. Duffy rubbed her eyes on her apron, straightened up, and set the kettle down with a bang.

"Mr. Duffy, do you recall Mr. Lincoln?" asked Mrs. Duffy.

"Are ye daft, woman? He was a great man. How could I forget? I remember I was just a nipper when that dirty actor shot him. I was learning how to milk then and my pa..."

"And do you recall what his great work was?" she continued. "The thing that all will hold in memory? The great kindness?"

"What are ye on about? He won the war against the rebels."

"And what else, may I ask? At the end of the war. It starts with E."

"Emancipation, of course. Are you taking up school teaching, then, putting me through my lessons?"

"I just thought to say that he left out a great large group of people when he freed the slaves."

"What on earth?"

"Yes, he left out the wives. We work all day for no pay. We take orders from others. Sometimes we are struck..."

"I've never struck you, Lavinia. Thought to, but I held back..."

"And we have no say in things, no vote, no property."

"Where have all these ideas come from—to disturb the peace in our home?"

"Truth will out, Henry, truth will out. I can read a newspaper as well as you—or better. I can think. I have plenty of time to think as I labor with these blasted pies. Here we are at the end of a long workday, my arms in the dirty water and you with your feet up on the stove. It doesn't seem fair. Can't you at least dry a few pans for me? My back is giving out."

"It's woman's work. It wouldn't be fitting."

"Is the sheriff going to come and investigate? You can dry a pan or two and no one will know. Inga won't tell."

The rocker creaked as Mr. Duffy stepped on a loose board in the wood floor. "Just this once, in honor of Mr. Lincoln, but I don't plan to make a habit of it."

"Ah, you're a great man, Henry, no matter what they say."

"Say? What do they say?"

Mrs. Duffy just smiled and handed him a towel.

Chapter 6

A Great Bargain

"Make way, Lavinia," Henry Duffy shouted one afternoon. "The flour's come."

"What flour?" Mrs. Duffy looked around, puzzled. "I didn't order any flour."

A burly man unloaded ten 50-pound cloth sacks of flour from his wagon onto a dolly and trundled them in through the door. "Where to, Ma'am?" he asked.

"Just drop 'em any place here," Mr. Duffy answered, "and thankee kindly for the transport from the train."

The man held out his hand hoping for a tip and Mr. Duffy shook it firmly. "Much obliged," he said, "Much obliged," as he herded the man toward the door and out to the wagon.

"My stars, Henry," Mrs. Duffy exclaimed as he came in and closed the door against the fall draft.

"Why ever did you buy so much flour? I make pies. I don't feed the Army of the Potomac."

"I got a wonderful price, Lavinia. Think of it. You always pay $2 for a fifty-pound sack, right? So that's" – holding up fingers and mumbling—"$20 for five hundred pounds. 'Course I could've got a better deal by the barrel. They give 194 pounds for only $3."

"Right. Your sums sound right so far. And I am glad you got it in sacks. They're so useful for making towels and clothes."

"Well I got these for only $1.35 each! What a killing deal! And it's the best red wheat from South Dakota on top of it, straight from the mill in Minneapolis. They practically gave it to us."

"I can't use that much flour in a month of Sundays, Henry, and it's blocking up my kitchen. I can scarcely move. We'll have to get it into the loft somehow."

"I'll get around to it, Lavinia--once my lumbago lets up."

Inga, coming in from school, climbed carefully over one stack of flour sacks and set down her strap of books. "Why is so much flour here?" she asked. "Do we have to make so many pies?"

"All in good time, my girl," Mr. Duffy said. "All in good time."

"Henry got a bargain," Mrs. Duffy replied. "A wonderful saving. Six dollars and fifty cents saved. Now we may be able to buy a bigger house, or even a farm." She didn't sound very happy.

"You do not like it?" Inga asked. "Flour is good."

"But not in wagonloads where I am trying to work," said Mrs. Duffy. "We'll be climbing over it for weeks while Mr. Duffy rests his back."

"I'd think a little appreciation is in order," Mr. Duffy grumbled. "A body can't get any respect, no matter how hard he tries."

All the rest of the week, Mrs. Duffy and Inga shuttled around the piles of flour, handing each other pans and rolling pins as needed to save the roundabout trip. Every bit of work took longer and by the end of the day everyone was grumpy and tired.

By the next Tuesday evening Inga was so weary of climbing around flour sacks that she decided to solve the problem herself. While Mr. Duffy was at the general store gossiping with his friends around the stove, and Mrs. Duffy was at a meeting of the Daughters of St. Patrick sewing circle, Inga executed her plan. If pulleys could lift the anvil, they could lift flour bags. She went into the shop and found a big strong screw hook, a long screwdriver, a length of rope, a brace and bit, and the pulleys. She hauled all her supplies and a stool into the loft. She left the extra rope in the kitchen near the flour sacks.

In the loft, she climbed up on the stool until she could reach the roof beam with the brace and bit. It was tricky standing on tiptoe on the stool but she managed to drill a hole in the big wooden beam. It would have been nice to have it right over the opening, but that was impossible. She pushed the hook into the hole until it grabbed the wood, put the screwdriver across it like a handle, and screwed the hook firmly into the beam.

After that it was just like lifting the anvil. Little by little she eased the block and tackle down, climbing carefully down and ladder as she did it. She tied a bag

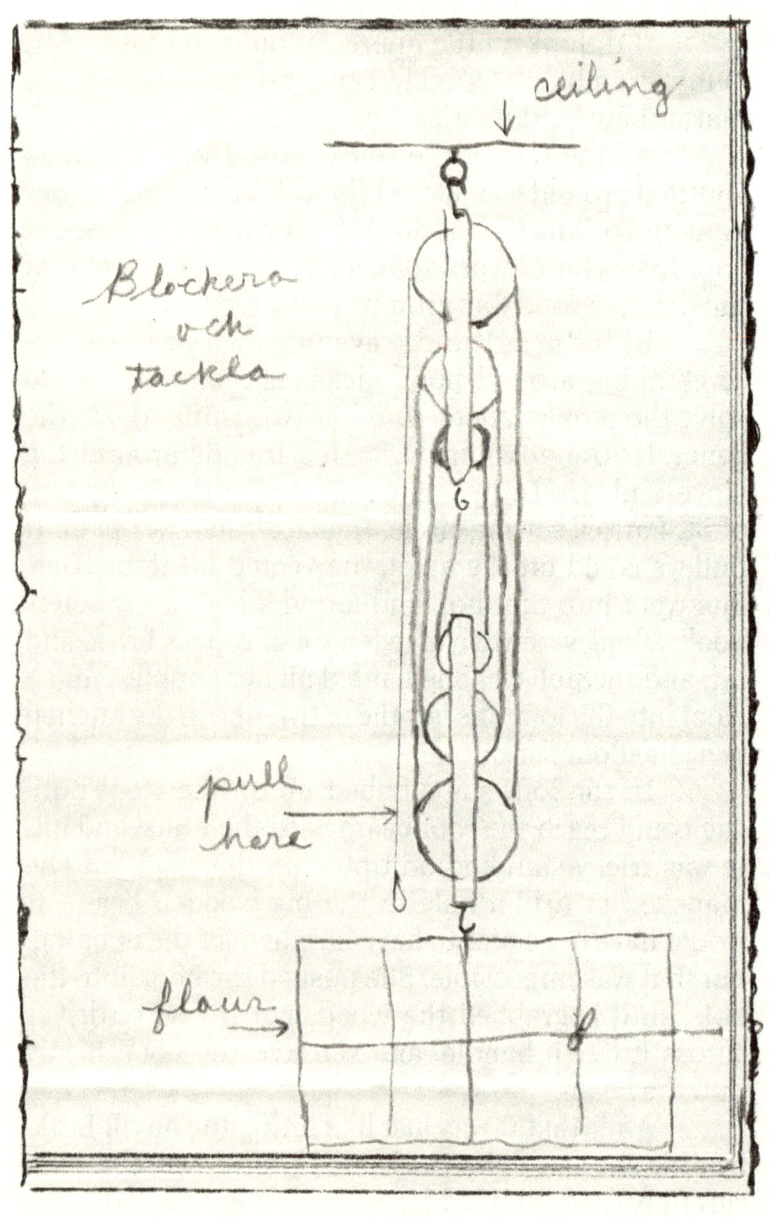
ceiling
Blockera
och
tackla
pull
here
flour

of flour securely with the extra rope, secured it on the hook and pulled it up into the loft. Then she went up the ladder, untied the flour sack and climbed down to repeat.

When she had three bags up in the loft the door opened and Mrs. Duffy walked in. "Inga, where are you?"

"Just working here, Mrs. Duffy. Come and see."

Mrs. Duffy turned back from hanging her coat on a peg and saw Inga backing down the ladder with the block and tackle in her hand. Some of the flour was definitely missing. "What in the name of sense...?" she sputtered.

"Moving flour up. It is like anvil. I can do this. See, I have three in loft."

"Seven to go and Mrs. Kelly will be throwing the men out soon. Let me help you," Mrs. Duffy offered. "I can tie them while you pull them up."

"In loft it is hard to move them."

"I will do that then." Mrs. Duffy reached down between her feet and grabbed the back of her dress by the hem. She pulled it forward up to her waist in the front and tucked it firmly into her waistband. "Now I can work," she said. "You've seen women's drawers before, I imagine."

Inga was shocked to see how quickly Mrs. Duffy climbed up the ladder with her black-stockinged legs and long drawers exposed to view. There was no time to lose, though, and the least of the problem was underwear. The conversation at the general store would soon be coming to an end. Inga tied and hooked the flour bags, then pulled them up to the loft. Mrs. Duffy untied them, wrestled them into place and sent the block and tackle back down. Mrs. Duffy was getting the last bag into place when Mr. Duffy came in through the shop.

"Hello. I'm home," he called and then caught sight of his wife at the top of the ladder with her black-stockinged legs and long drawers showing. "What in tarnation?"

"Just saving you work, Henry," laughed Mrs. Duffy. "Inga's anvil idea worked for flour too."

"I thought someone had been in my shop. The long rope was missing." He looked at Inga, carefully winding up the block and tackle. She coiled the rope and held it out to him. "Here's your rope," Inga said. She loaded the long screwdriver and the block and tackle into her outstretched apron. "I take all tools back now." Inga marched back to the shop.

"You did it yourselves! What good thinking. My sore back thanks you. Getting that girl was the best thing we ever did." He held out his hand to Mrs. Duffy who had fluffed her skirts properly and was stepping calmly down the ladder. "I have to admit the two of you together are pretty clever."

Chapter 7

Reading, Reading, Reading

A few days later at school, Inga finished the 4th McGuffey's reader. "Now I'm finished," she told Emmet. "I know all the words. I don't need any more."

"But what about the 5th and 6th? There are lots more readers and lots more words. In the 5th, here, there's a great story about the King of Prussia."

"I don't care for kings."

"It has a happy ending. A girl learns to read and then she saves somebody." He opened his book to page 42 and read: " 'How much happiness was Ernestine the means of bestowing through her good elocution, united to the happy circumstance that brought it to the knowledge of the King...' "

"Too many words. I told you. I don't care about kings and so many words. I'm weary of words, words every day. That's all we do. If we read about something to make, will Sister let us make it?"

"There's nothing about that in McGuffey's. You have to look in encyclopedias or *Scientific American*."

"Like an experiment?" Tom, overhearing, sat up straight, all ears. "We could ask if we can make a demonstration. We did that at Miss Kelly's school. Maybe Sister will give us leave."

At recess they found Sister supervising jump rope. "We have an idea, S'ter," Emmet began. "Could we make something that teaches us? Something that we read about first?"

"That would be most unusual, Emmet. This is not a shop."

"Maybe something from history," he persisted. "Like a Roman chariot, or a guillotine? Something educational. Maybe a model castle. Just a thing we can touch, to go with all the words?"

"Sister Ludwig has told me about your educational inventions. I have grave doubts about the value of these demonstrations. And last year she was almost injured by one of your experiments, as you well know. It would have to be something harmless and safe you can build during recess. Once it is finished you need to write a report and read it out to the class. No nonsense."

"Oh no, S'ter! No nonsense. Inga will help us. She doesn't do nonsense."

"What means 'nonsense'?" Inga asked.

"It's hard to explain," Tom whispered softly, "but you'll know it when you see it, or Sister will tell you."

Sister rubbed her chin thoughtfully. "Perhaps you can find an idea in a book we have here in the press. Relate it to a famous person we study. A whole chapter on the Renaissance Period in Italy is coming up. Maybe you can find something classical."

The next day Emmet, Tom and Inga spent their recess time exploring the Renaissance section in the encyclopedia. "There was really a lot going on then," Emmet said. "Wars, diseases, plagues, kings losing their thrones."

"I find a pope here with many children," Inga offered. "Would Sister like that?"

"Popes with kids will probably aggravate Sister." Tom laughed. "We better stick to something we can build. Didn't they used to have any inventors or explosions?"

They searched on, huddled over the book, until they found a page about Galileo. "What about Galileo?" marveled Inga, looking at the pictures. "He has so many moving things. Balls rolling, things dropping off towers, things for looking at the sky. He was very busy."

"Let's do this one," Emmet crowed, pointing at an illustration showing Galileo among a group of scholars. "We can get different kinds of heavy balls and roll them down. I wonder where we can get cannonballs."

"We must not do a war," Inga said. "Cannonballs can kill."

"Only if they are shot out of a cannon. They can't do anything by themselves."

"Maybe they have some left over from the war," Tom suggested. "Maybe at Gettysburg, but that's too far away. We can't go to Virginia."

"That's Pennsylvania, Tom. You don't know anything. Virginia had Manassas. Maryland had

Antietam. You have to keep 'em straight."

Inga remembered the long ride on the train. "Aren't they all near New York? It's too far, and how would we carry them all back here?"

"How about billiard balls?" Tom suggested. "McNally has them over at the saloon. My brother Pat went there once and played. When Mother found out she gave him a whacking! He never went back. Or," he paused. "I don't think he did. Mother didn't catch him anyway."

"How big are they? What is billiard?"

"It's a game the men play," Tom explained. "They have all these different colored balls laid out on a table. The balls are about as big as your fist." She nodded. "They poke them with long sticks to make them fall into holes at the corners."

"That's stupid," Emmet said. "Why don't they just put them where they want them and be done with it?"

"Well, the men like it. They drink beer, argue and win money. My brother said it was fun."

"They must not know about inventions, then. Say, maybe McNally would loan us billiard balls to do the tests like this Galileo."

"We can build a nice path of wood," said Inga. "Look, this thing in the picture Galileo made for a king. We can make that. The balls roll down. We see which one is faster. Where can we get wood, Emmet?"

It certainly looked easy. The picture showed the king and some priests watching Galileo roll balls down a smooth ramp about ten feet long. It was held up at one end by a fancy post. "Can't be much trouble," said Emmet. "No sparks, no danger."

"No explosions either, probably. It sounds dull," said Tom.

"Galileo just sets the balls on the board and lets them go. It doesn't look like much," Emmet agreed. "I wonder why they made a fuss about it, writing in the encyclopedia and all."

"We could make it very smooth," Inga said. "Use different kinds of balls all the same size. Just see what they do. Maybe some would be faster."

"Might as well," said Tom. "Even if it's dull, it's better than McGuffey's stories. Maybe my mother has a board at the store we could borrow."

"We can use books for the tower," said Inga. "Many books are there in the corner."

The next morning as the school bell was ringing, Tom and Emmet struggled through the classroom door lugging a twelve-foot plank. Inga followed them carrying a cloth sack filled with heavy colorful balls. "Careful with those balls, Inga," Tom said. "When his pa gave them to me yesterday, Eamon made a big fuss that we might lose some out of the set. Said they were special, ordered from New York, made from special stuff like ivory or something. That Eamon is so stuck up. He thinks experiments are stupid and said we're stupid too."

Sister's jaw dropped open when she saw the size of the plank. "It's our experiment," Emmet announced. "It is completely safe. No sparks. No explosions. And it is copied from a classical person called Galileo. It was in your own encyclopedia."

Sister Norah James heaved a great sigh. Sister Ludwig had warned her that Emmet and Tom were very creative and loved to try out their ideas. It is more of a challenge than I need at my age, she thought. I can't correct them; they know all the answers. Sometimes they have questions I can't even answer. And now we're to be treated to the sight of balls rolling down a plank. We set out to help the new girl learn English. I don't see how a 12-foot long plank accomplishes that.

"You must be very careful with this, children. And you boys must help Inga to have all the words she needs to do it. First things first. Put your supplies up behind my desk and you may work on it during recess."

"Thank you S'ter," they said together. "You won't be sorry."

"I certainly hope not," she said.

Chapter 8
The Ramp

The next day the board and bag of balls didn't look like much in the early winter sun that shone through the classroom window. As they shifted the heavy plank around, the dust behind Sister's desk rose up in little puffs. "We don't have room. We must move desks," said Inga. "I will take screws out from floor. Move and put screws in again."

"Sister'll have a conniption fit if we move her desks. We'll get in trouble."

"Maybe she won't notice if we do it when she's out," suggested Emmet. "We'll only move the front row to the back. It will look the same."

Tom shook his head doubtfully, but Inga was already at work with her screwdriver. By the time the early bell rang and the children streamed in, there was extra room at the front of the class and the screwdriver was in Inga's pocket. As Sister Norah James neared her desk, she slowed down, looking around curiously. Something was different, but what? After a moment, she noticed twelve neat holes in the wooden floor where the six desks used to be. She pulled a handkerchief from her sleeve, took off her spectacles and polished them vigorously. She stretched out her arm toward the students' desks. "They're farther away," she mumbled to herself, shaking her head.

The children settled in to their usual places. The desktops looked unfamiliar. The scratches weren't the ones they were used to. A couple of desks had empty holes where the inkwells had been. A puzzled murmur spread through the classroom and students looked at each other. Finally Emmet put up his hand. "We shifted the desks to make way for our project. Nothing is hurt. We can put them back when the experiment is done. You don't mind, do you, Sister?" He smiled his sweetest smile and crossed his fingers behind his back. "Nothing is broken. See, all the desks still work the same. It will be very educational."

Sister scuffed at the screw holes with the tip of her shoe. In all her years of teaching, she had never faced such a situation and was temporarily speechless. "When do you plan to build this educational object, Emmet?"

"We'll start at recess. It should only take a few days. Then we'll take it apart and put the desks back."

Sister thought about it and couldn't come up with a good reason to refuse. "Very well. By the end of next week I expect to have everything back to normal. Is that understood?"

"Yes. We'll do it." Inga started to clap her hands, then stopped herself and put her hands under her apron. She looked around to see if anyone had seen her clapping but nobody was looking. They settled down to their bookwork. Recess couldn't come too soon.

Once the children shrugged into jackets and streamed out for recess, Tom and Emmet laid the plank on the floor in front of Sister's desk. "It looks like a dumb idea already," Tom said. "Nothing's happening."

"One end must be up," said Inga. "Balls will go down. If the end is high, they will go fast."

"Let's measure how high and how fast!" Emmet pulled out his worn notebook. "Get sister's ruler and see how long the board is."

"Mother told you, Emmet. It's a twelve foot 2x10 joist," Tom said. "She ought to know."

"Scientists always measure. You can't just go by what people say."

Inga got Sister's ruler and measured end to end, turning the ruler over and over like an inchworm. "Eleven feet, eleven inches."

"See, I told you." Emmet recorded the length in his book. "Now we need to prop it up. In the book they had a nice pillar."

"Galileo was working for a king. We don't have any pillars."

"We can use rocks or desk or boxes. No difference." Inga marched to a stack of old books behind the desk. "Let's use books."

She started to move the textbooks. They were too old and tattered to read any more, but too good to throw away. Tom helped her stack them.

"Should we use books or inches?" Emmet

asked. "We have to decide how high to go."

"Books," said Tom.

"Inches," said Inga. "If we count books, then if somebody tries to copy and they have bricks, they will get mixed up."

"OK." Emmet started stacking. "Should we measure from top or bottom of the board? It is thick." This required a conference and they finally decided to measure from the top of the board every time because the balls would always be on top. Three books gave them a five-inch height and Emmet recorded it just in time to be in his seat as the kids came back.

Somebody shouted, "Ya buildin' a barn in here or what?" There was lots of shushing. "Maybe a bridge?" More shushing, then Sister came in, took one look, pressed her lips tightly and turned to the class. "This is a project for our Renaissance studies, children. There will be no rude comments to fellow classmates. It may prove to be more interesting than you expect," Sister said, mostly to reassure herself.

Every day at recess they added more books to the stack, measured the height, and rolled a ball down to see how fast it would go. Emmet recorded all the measurements. "Something is strange," Inga noticed. "The ball goes faster and faster as it gets down near the bottom. Is it supposed to do that?"

"It goes down in a blink," Emmet crowed. "Too fast for even 2 ticks of the clock."

"That's no science," Tom said. "Everybody knows things roll down steep hills. And they speed up too. Everybody knows that. Galileo couldn't be famous for that!"

They went back to the article in the encyclopedia.

"Says here he tried different kinds of balls. What can we get that's the same size as the billiard balls, but not so heavy?"

"Remember? I said that idea," Inga reminded them.

They tried whatever balls they could borrow: baseballs, wood balls, balls from games, homemade balls. The best turned out to be the billiard balls after all—smooth, equal in size and weight. The colors didn't change the results. Every day they sanded their board to take off bumps and slivers. Inga convinced the boys to only change one thing at a time and the easiest thing to change was the slant of the board.

They found out what they already knew—a big slant meant more speed. The stack of books held the end of the board up about 4 feet; the balls rolled down in 2 ticks of the clock. But Tom wanted speed. "How high do we need to go so the ball gets down even before the first tick?" he wanted to know. "How high do we go so it's just an instant? A zero time?"

They agreed that it would have to be pretty high, at least another couple of feet. The old books were all used up and Sister wouldn't let them use good ones. "I have all my old *Scientific American* copies saved, ever since I was a kid, almost three years' worth. I'll bring them tomorrow," Emmet said.

Next morning Emmet staggered in, his arms loaded with old *Scientific Americans*. The copies were well-used and slid every which way as he walked. As soon as the recess bell rang, the team hurried to add them to the top of the stack. As the stack grew to five feet, the top layer began to teeter. "It has to be straight," said Inga. She adjusted the magazine pile so that the sides were even. "If the board presses down uneven, it will fall."

"That would be something to see!" Tom said. "Almost as good as an explosion."

"Emmet reminded him, "We promised. Remember?"

Finally when they got to six feet they ran out of magazines. The pile leaned over in a few places but a slant on one side pretty well matched a slant on the other side. "That's it," Inga said. "Let's put the board up very gently and hope it stays."

It stayed, as long as nothing else moved. Inga climbed up a chair and reached out with a ball. "Watch the clock!" They watched and on the tick she let the ball go. It hit the floor just past the next tick. "Just over a second! That's the fastest ever." They danced around with joy. The magazine stack wobbled—and then settled back into place. "We'd better keep them all away," Inga said. "Just carefully set the bag of balls here next to it. Can we put a sign up to stay away?"

"I'll do it," Emmet said. "I'm good at signs." He wrote DON'T TOUCH in big letters on a slate and propped it up on Sister's desk. He wrote it again on the chalkboard. "That should do it." The sound of twenty-six pairs of boots shuffling in the hall startled them. "Let's stand at the door and warn everybody when they come in," Tom said quickly.

As the class entered, they were surprised to see Sister Ludwig walk in with them. "I'll be teaching for the rest of the day," she announced. "Sister Norah James has a bad toothache."

Emmet looked at Tom. "Remember the electric experiment last year?"

"I sure do. She got knocked flying."

Emmet tugged at Sister's sleeve as she went by. "Excuse me, S'ter. Be careful of the science experiment up near the desk,"

"Of course, Emmet. Now go to your seat." She turned to the class. "Quiet, children. Take out your slates for penmanship practice."

Once they settled, she began again. "I want to show you the proper way to form your capital Q's. Now where's a slate?" She swung around, her sleeve grazing the tottering pile of books. She erased Emmet's message with the heel of her hand, drew a large Q on the slate and held it up. "See, it's quite like the number 2 but you need a graceful loop at the top. Let's all try it together."

She turned and stubbed her toe on the bag of billiard balls. "Ouch. Now why would someone leave a bag of balls on the floor here?" she asked as she set down the slate and picked up the bag of balls. The magazine pile wobbled. It did not fall. Inga let out a deep breath. "Don't move them," Emmet said, "they go with that pile of magazines."

"Why, then keep them together," she said as she plopped the bag of balls onto the stack of magazines and turned back to face the students. "Now, be sure to make your loop graceful," she said holding up her slate again. But the children could not concentrate on their Q's. They were staring at the magazine pile, which has started to wobble. The starched sides of Sister's headdress were like blinders and she couldn't see what was happening, but the children could. Frozen in shock, their mouths gaped. Slowly at first, then faster, magazines started to slip out. The board tipped and the bag fell open. Balls began to drop, bumping each other and bouncing along the floor. "What's that racket?" Sister Ludwig said. She turned and grabbed at the board to steady it which knocked over the stack of old books. A ball dropped onto the top of her foot, smashing her sore toe. "OWWWW!" she cried. She fell on her rump, hard. The class gasped, and then some of the boys started throwing the balls to each other. Billiard balls are heavy and hard and soon there were cries of "Hey, stop that" and "That hurts." Sister sat slowly shaking her head as she adjusted her veil.

"That was a great demonstration of gravity, S'ter. Better than Galileo's," Emmet said. "But didn't you see the sign? It's important to follow directions, you know. I wrote it up big on purpose."

Sister looked up at the concerned faces of Inga, Tom and Emmet and said to Inga, "I hope you can bring some sense into these new friends of yours. They're trouble, just waiting to happen. Maybe as a girl you'll have better sense."

It didn't work for you, though, S'ter, Tom thought.

Chapter 9
The New Old Dress

"That wind is getting really chilly. I want to get the washing done tomorrow before freezing sets in," Mrs. Duffy remarked the next Monday. "That dress could use a wash, Inga. You've been working in it every day, in the shop and in the kitchen. Go change into something else."

"I tell you before, Mrs. Duffy. Remember? For church? I have no more."

"What were they thinking to send you across country with only one dress? Your parents must be either very poor or completely thoughtless. You can't be going about in your shift every time we wash. We must get you another dress so I can wash the soiled one. Let's go to Kelly's General Store. Kate Kelly sells everything. She may have a dress, and maybe a coat too. Your jacket won't stand up to an Iowa winter. Let's go right now." She glanced up at the clock. "The pies have almost three-quarters of an hour. We can make it back in time."

Mrs. Duffy tucked her hat down over her hair and stabbed a long pin through it to hold it in place. She buttoned up her coat. Inga followed her, trotting behind as she pulled her shawl over her thin jacket, hoping to cover the open places at the front where the buttons wouldn't reach. They walked quickly down the street, narrowly avoiding Tom's little brother who zoomed around the corner, chasing a skinny cat. Why does he chase it, Inga wondered. It only wants a home.

Mary Kelly Ronan was just sweeping the store when she heard her bell jingle as they walked in. "Morning, Lavinia," she said. "My first customer. You are out very early on this crisp morning."

"I'm surprised to find you here, Mary. Where's your mother? I thought you'd be out at your farm in your own home now that you're married."

"Oh I'll be back at the farm soon enough. My mother was called to tend her Aunt Moira down to Cedar Rapids. She's getting on in years and came down with a bad chest this fall. I'm watching the store and my brothers while she's gone. How can I help you?"

"We have a problem, Mary. This is Inga. She came to us on the Orphan Train a month ago with just the clothes she stands in."

"Yes. Tom told me about her. He says she's smart."

Mrs. Duffy turned to Inga. "This is Mrs. Ronan. She used to be a teacher last year before she married."

"Are you Miss Mary? Emmet says you had a good school."

"Yes, I was Miss Mary last year. Now I am Mrs.

Ronan and I live on a farm with my husband Jim. You must know him. He brings his horse in for shoes."

"She needs clothing that will take her through the winter," Mrs. Duffy interrupted. "Can we hurry? We have pies in the oven and there's no time for a chat."

"I have a nightgown," Inga whispered, "and stockings."

"Yes, yes, but a dress and coat that fit are needed. If we have a repeat of last winter I don't know how we'll stand it. What do you have that will fit her, Mary?"

"I do have two dresses, but I fear they are out of your range, Mrs. Duffy. Mother ordered them for the Christmas party trade. You may as well have a look." She reached up to a high shelf and took down a box. As she opened it, Inga gasped. One was pink and one was light yellow, smooth and silky with ribbons at the top. Mrs. Ronan shook them out and the ruffles at the hems floated over the counter. Inga had never seen anything so beautiful. Then she thought about her life in the kitchen making pies and filling the stove. She started to open her mouth, but before she could get a word out, Mrs. Duffy shook her head. "These aren't for life in a blacksmith's family," she said. "Don't you have anything for work? Something warm and sturdy that will wash well?"

"Nothing new. Do you mind second hand?" Mary asked.

"That would do," Mrs. Duffy said over Inga's head. "She's just an orphan after all. She doesn't expect too much." She craned her neck to see the store clock. Fifteen minutes gone.

Mary called to her sister in the back of the store. "Katie, get those two dresses you outgrew and your old coat from the attic, please, and bring them down here. We may have a use for them."

Shortly Katie came down, her flyaway red hair falling out of its pins and her arms loaded with clothing. "She just got a spurt of growth last year, Mrs. Duffy," Mary said. "And what with her going into long skirts now, she couldn't wear these any more. She's not hard on her clothes though. All the buttons are in place and nothing is torn. My sister is careful of her things."

Katie plopped the items on the counter: a brown dress with little black buttons down the front, one of sprigged green cotton with rickrack at the neck, and a dark blue wool coat worn shiny at the back from sitting. Then she went to the back of the store and resumed straightening the boxes on the shelves.

Mary held the dresses up against Inga. They were long and wide, but free of spots. "She'll grow into them, Mrs. Duffy," she said. "Turn around, child." Inga did. Mary held the coat up to her back. "If it's long it will protect her legs better. You'll only have to sew up the sleeves."

Maybe the sleeves will protect my hands too, Inga thought. Or I could tie them together in the front like a muff. She giggled to think of her long sleeves flapping like wings in the wind.

"We'll take them, if you're offering," Mrs. Duffy said. "How much do they cost?"

"I'll trade for a couple of pies. We have no more girls and my wild brothers will have no use for them. The clothes will only feed the moths in the cupboard.

Here, take this scarf too—it goes with the coat."

"I do not beg," Inga heard herself blurt out.

"Why, Inga! Mrs. Ronan is doing you a kindness."

"I pay. I will work for clothes. How many days, Mrs. Ronan?"

Mary took a deep breath, then smiled. "Make me an offer," she replied. "How many afternoons are the warm clothes worth to you?"

"I can work on Saturdays after work, until Christmas, if Mrs. Duffy lets me."

Mrs. Duffy nodded.

"You have a bargain then. I like to do business with an upstanding person like yourself, Inga. Come after lunch Saturday." She held out her hand to the child, who had a surprisingly strong grip, and looked her in the eyes for the first time. "Welcome to your new job at Kelly's Store."

As they hurried back home, Inga inquired, "Are these fripperies, Mrs. Duffy? What will Mr. Duffy say?"

"Lordy, no, child. A person needs clothes and a growing child needs them more than anyone. Mr. Duffy only grows around his belly. They don't call them 'coveralls' for nothing. He doesn't need new things so often." She walked a few more steps. "You needn't tell him what I said, Inga. About growing at the belt. He wouldn't want to hear about that."

"I won't. But I know what you mean." Mrs. Duffy returned her sly grin.

Once they had returned home and pulled the slightly overdone pies from the oven, Mrs. Duffy dug her sewing basket out of the chest near the bed. It was loaded with scraps, rolls of different-colored thread, a pack of needles, a red cloth pincushion that looked like a lopsided tomato, and two scissors. "Here, Inga. Thread your needle and sew up these dress hems, about 6 inches or so." She folded one of them up about as wide as her spread-out hand would reach. "You do know how to sew, don't you?"

"Yes. My mother teach me before school in Sweden. Every girl knows how to sew. Mother had fast hands. She told me, 'If you need a helping hand, look at the end of your own arm.'" Inga pulled out a needle, threaded it with brown thread, licked the end of the thread, wound it around her finger into a knot and held out her hand for the brown dress. "I start with the warm dress, Mrs. Duffy. I like the pretty buttons."

They sewed together for a while by the dim autumn light shining through the window near the stove. Mrs. Duffy cleared her throat and slowed her stitching. "Why ever did your father send you off in such a hurry, Inga? Not even a proper change of clothes to go off across country to a new life. I've been puzzling on this ever since you came to us. You are a good girl, smart, hard-working, clean about her person, obedient—how could he send you off? What happened? He is alive, isn't he?"

"Yes. He lives at a boarding house, Mrs. Lundgren's in New York. He has his own room like the other men. Someday he will write to me."

"And you lived there with him?"

"Yes."

"Who took care of you?"

"Mrs. Lundgren. She took me to school sometimes, to American school."

"And on the other days? Did you help her?"

"I helped her when she was weak. All the men need supper and she is weak from *schnapps.*"

"What is that?"

"She drinks it. I tasted it once. It was bad. It makes her weak. She sleeps. It makes the men happy, though. They sing and laugh. Sometimes I was scared."

"Were they mean to you?"

"Oh no, they loved me. They always wanted to pet my hair and hug me. Then one day my father came early home from work. He saw Mr. Larson petting my hair and hugging me when I am putting out plates. He hit Mr. Larson down to the floor and shouted 'Leave my *dotter* alone.' Then he said to me 'I must find a better home for you. This is not good.' Next day he threw my things in a carpetbag and took me to Children's Aid. He told the lady I need a good home. I can't stay anymore with him at the rooming house of Mrs. Lundgren."

"Just like that? No explanation?"

"The lady said 'You live in a boarding house?' He looked a funny way at the lady and she nodded her head. 'We will find a proper home for her. Some place with a God-fearing family, a family with a mother. We have a space on the train going west this afternoon.' She gave the papers and father writes his name many times."

"What happened then?"

"Father said to me, 'Be brave. Be like a Viking. You are a good girl. They will be lucky to have you. Mother would want this for you if she lived.' Then he hugged me and put my hand in Mrs. Hobson's hand. He left the office. I thought he would come back for me, but we went to the train and he didn't come. Here I am. I must make the best of it." She brushed away a tear with the back of her hand.

"...if she lived? Didn't she stay behind in Sweden?"

"She is gone. I have only father, and he gave me away." She stabbed at the cloth with her needle and sniffed.

"Do you think he will come for you someday? Can't he at least write?"

"Maybe he can write. He went to school in Sweden. He is very busy being engineer. He has no time for small things like writing."

"But he should. You are his child."

"I cannot think of it. I will cry and then I won't be able to work. I don't think of it. I will write to him again soon. Maybe he will write back. I hope for that."

Mrs. Duffy held out her arms and wrapped Inga in a great hug. "You will be my daughter, even if it's only a few years. You can call me Ma Duffy if you want to. Your father was right. We are lucky to have you. Now let's get these clothes sewed up before Mr. Duffy comes in hungry as a bear."

Chapter 10

Schooling Comes Second

"No school today, Inga," Mrs. Duffy announced. "We have a load of washing to get done before cold weather sets in. We can put your gingham dress in with Mr. Duffy's overalls. It's dark so the color from his filthy pants won't matter."

"There is no school? I hoped to finish my sums and start reading the 5th Reader."

"You know school isn't your main concern, Inga. You're here to help me. If there's free time, then you go to school. It is a working day for you my girl. This is no job for one person." Inga nodded, picked up a bucket and followed Mrs. Duffy out of the house, glad for the warmth of Katie's old dress.

In back, near the privy and the shed, Mrs. Duffy had set up a bench with two big wooden tubs and a contraption with rollers in between. A big iron pot with a spigot stood on a rusty potbelly stove. "Fill the pot with water first, Inga," Mrs. Duffy ordered as she laid a fire in the stove. "We'll need to heat water for the white things—the towels and underwear, the sheets and blouses. I want to boil these longjohns well."

"Yes, I know. I helped my mother when I was small. Should I cut up soap for the hot water tub?" she said, taking her folding knife out of her apron pocket.

"Yes, dear. That would help." Inga pumped three buckets of water and poured them into the pot. "Now we'll need cold for rinsing. Just put it in this second tub." Inga did it.

"I like this," Inga said, pointing to the thing with rollers. "We don't have this in Sweden."

"Mr. Duffy bought this for me, bless his heart. It's called a wringer. It's the latest model. The soapy clothes go through here, right into the cool rinse water," she said, pointing between the rollers. "Then you turn this handle." She gave it a half turn. "When you crank the rollers in reverse, the rinsed clothes come through. The water is squeezed out of them, ready to rinse again or hang up to dry. I wouldn't know what to do without it any more. It is such a help."

They worked through the clothes in batches, starting with whites. Mrs. Duffy scrubbed each garment on her washboard, lingering on any spots. Socks and longjohns were a particular challenge.

Inga worked fast, maybe too fast, cranking the handle like a whirlwind. Three socks and a shirt landed back in the soap.

"You have to catch them before they fall back into the soap," Mrs. Duffy warned. "And watch your fingers!"

Inga turned the handle, slowed her speed

and soon they were working like a human machine, squeezing each item into the rinse water, then back, wringing out the rinse water. Soon the clean damp clothes filled a big basket, ready to hang. They hung them on the rope that ran from the house to the privy, and spread the sheets over the bare bushes.

"Well, that's a big job done," said Mrs. Duffy. "If we're careful we won't have much besides kitchen towels until spring. I hate to do washing in winter, having clothes flapping in my face all day when I'm trying to bake. It's lucky I don't have a baby. Those diapers never end." She started toward the house. "Dip the water out and pour it around the bases of the bushes and the apple tree," she said, "I'm going to rest while the laundry dries. Come in yourself after you throw away all the water and douse the fire."

Inga used a pitcher to dip out the cold water and carried it to the bushes. I wonder why she doesn't just put a hole in the bottom of the tub and plug it with a cork, she thought. Or maybe put the washtub on top of the stove and heat it all at once. No, the wood might catch fire. As the ideas ran through her mind something else popped into her mind. If sheets and even long-legged underwear could be pressed flat, why not pie dough?

She slipped back into the house quietly. Mrs. Duffy was napping on the bed with her boots off and her hair undone. Inga took a ball of the waiting dough and tiptoed out the back. Maybe this will make a pie machine, she thought, a machine for two important jobs. She flattened the dough ball and tried to stick it between the rollers. It wouldn't go. The rollers were still wet from laundry and smelled of lye soap.

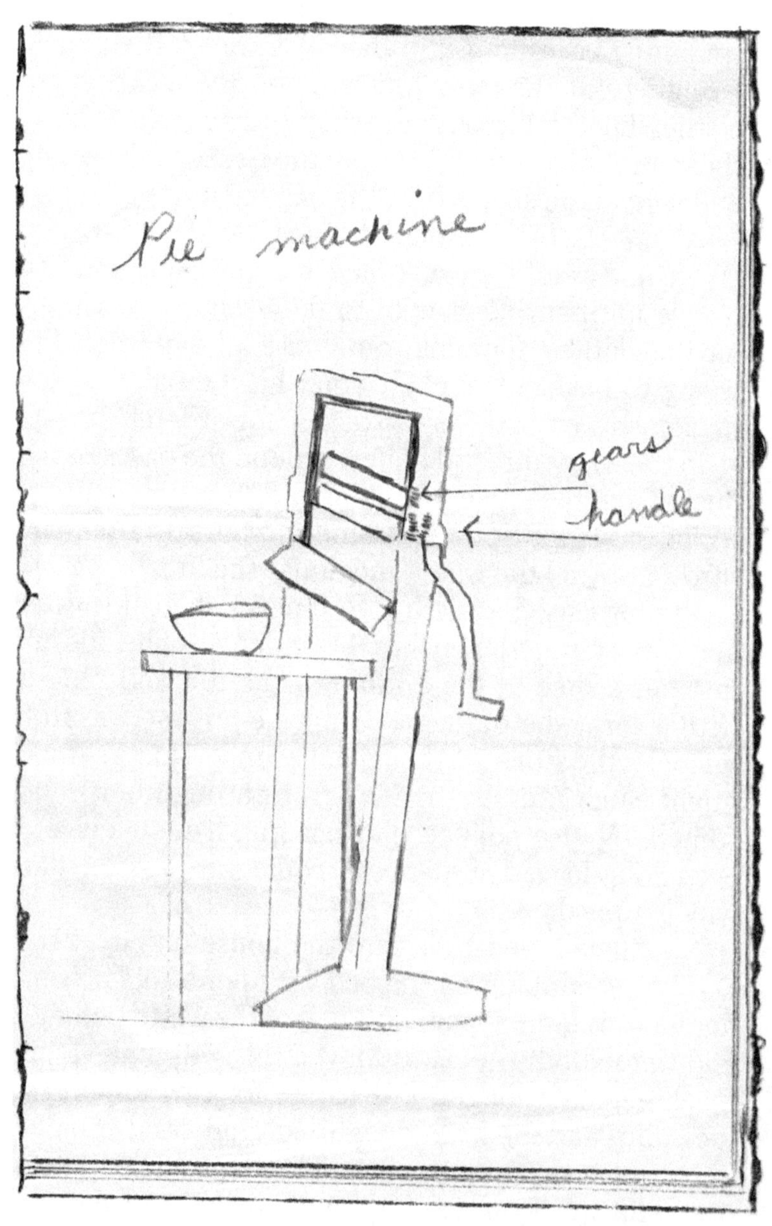
Pie machine
gears
handle

She flattened the dough some more with her hands, dropping pieces of it into the dirty soapy water. Soon she had a big flat piece of dough balanced on her hands, flopping down at the sides. She got the edge started between the rollers and turned the crank. Some of the dough went through. The rest sank bubbling into the dirty water. Maybe it will work better wet, she thought. She fished it out of the water to try again. This time much of the slippery dough went through but a lot of it still stuck to the rollers. The piece that went through was rubbery and smelled of lye soap.

"Haven't you finished, Inga?" Mrs. Duffy said, coming out and standing behind her shoulder. "I feel better after... What's happened here? Oh, my stars and garters! What a mess you've made! Whatever were you thinking?" She grabbed a freshly washed towel off the rope and started to scrub the rollers. The dough came off in greasy bits and stuck to the towel. The more she scrubbed the worse it smeared. She turned suddenly with her hand and the messy towel lifted, but froze. Inga was huddled down, covering her head with her hands and shaking. "I so sorry," she said. "Please don't send me away. I will clean everything. Please don't give me back to Mrs. Hobson."

Mrs. Duffy lowered her hand and sighed. "It's only dough," she said. "Did you think I'd hit you? I won't send you anywhere. This is your place." She put her other arm around Inga's shoulder. "But you have to help me clean up this mess."

"I wanted to make pie machine. You work so hard."

Mrs. Duffy just shook her head. "I see the idea, but you know this may be an invention that isn't ready to be invented yet. Give me a hand here." They scrubbed the rollers with the used soap water and Inga poured a bucket of clean water to rinse them.

"I don't think we need to share this story with anyone," Mrs. Duffy said. "We'll just consider it a brave attempt and admit failure. Let's hope the clothes dry before it rains. Anything frozen can finish drying near the stove inside."

"I am thinking," Inga remarked later. "When you make the pies, the roller is dry."

"Yes."

"And covered with flour."

"What of it?"

"Maybe if there were rollers, like the wringer, that were dry with flour, maybe it would work."

Mrs. Duffy considered this for a bit, gazing at the hopeful expression on Inga's face. "Perhaps we can try in winter, when the rollers have completely dried and we have time on our hands."

"We can bring the wringer in side?" Inga asked, brightening up. "Right here near the stove? Will you let me try?"

Mrs. Duffy sighed. "I will let you try. I will help you bring the wringer inside. Only when time hangs heavy, though. Work comes first." I must be getting soft, she thought to herself. No one in her right mind rolls pie dough with her washing equipment.

"Oh, Mrs. Duffy, you are the best," Inga cried. "We can work together to make a pie machine. Just think how it will make your work easy."

Or incredibly messy, Mrs. Duffy thought to herself. Time will tell.

Chapter 11

Working at the Store

Inga tucked her chapped hands into her coat sleeves as she hurried to work at Kelly's General Store the next Saturday. The five Saturday afternoons would pass quickly, she thought. It was only fair to pay for the dresses somehow, but time for reading in the loft was being lost. Books will wait, she told herself, they will wait. Clothes are not free. She scrunched her neck against the snow flurries and walked faster.

As the entrance door slammed behind her, she gave a long look around the store. She had been there with Mrs. Duffy before. However, at that time she had been too worried about burning the pies back home. What a wonder she saw now. On the high shelves were big rolls of cloth for sewing everything, cotton and batiste for baby dresses and serge for woolen suits. Boxes of tools and devices filled the space as far as her eyes could see. There were barrels filled with nails and things she didn't know the names of. The store was cluttered with wheels of all sizes, harnesses for horses, boots and overalls. On the east wall stood a shelf with more than twenty books, all new and clean. Next to it was another shelf, with a locked glass door, full of medicines of all kinds, ointments, salves, lozenges, tonics and tinctures. The other whole side of the store held shelves with sugar, salt, cinnamon in small packets and chocolate with a picture of a lady on the package. She stood for a moment, taking it all in, her mouth open in amazement.

"Please come and help me here," Mary Kelly Ronan called. "This can of barn paint is almost too much for me to lift." Inga ran to the back and helped hoist the five-gallon bucket onto a low shelf. "*Barn* paint? You don't paint children in America do you?"

"Why would you say that?"

"In Sweden a *barn* is child."

"In America a barn is where the animals live," said Mary. "Now let me show you what your work will be. You will be selling the items in the store. Come and see where we keep the money. It's called a cash drawer. The ledger is inside as well."

Inga's eyebrows shot up. "What is that?"

"A small book for business. When people buy, you write down how much they paid. You'll have to cipher out how much to give back in change. You can do your sums, can't you?"

"Yes."

"In your head?"

"Most of the time."

"You can write the problem on these paper scraps I keep under the money if you need to. It is important to give correct change."

"What if they don't give me enough money?" Inga wanted to know. "Some people get pies and pay later. I help Mrs. Duffy with that."

"Ah, yes—the credit problem. Some of our neighbors here are down on their luck from time to time, often just when they need seeds or flour or warm clothes. If this problem comes up, just call me. I know who is honest around here, and who would let the store wait until judgment day to get paid."

"Is that all I do?"

"You will help people find what they need. You will measure out cloth as needed and cut or tear it carefully. If there are no buyers you can work with Katie to neaten the shelves and put things back where they belong. Remember, keep smiling and use best manners."

Inga set out to be the best sales lady she could be, but her eye was always out for the tools. When it was quiet, she lined up all the screwdrivers by size in their box. She hung the saws neatly on the wall by size, from the two-handled lumber saws to small keyhole saws, so their shining sides and smooth varnished handles would show. Little by little she organized the tools to show their strange beauty. When Mrs. McNally came in to buy gingham for a daughter's school dress, she couldn't get over it. "I've been in many a fancy shop back before I married Mr. McNally, but I never saw such an array as this," she told Mary Kelly Ronan. "One doesn't expect a lovely display of tools like this out here on the prairie. Your mother will be proud when she returns."

"Oh, it's our orphan girl, Inga, who does it when

she's not busy," Mary replied. "She likes to handle the things and put them in order."

"You've done well, Inga," she said after Mrs. McNally left.

"Everyone knows you think better when your tools are in order," said Inga.

After that, shopping on Saturday afternoons picked up. It may have been because Christmas was coming, and often times Inga was the only available sales person. "I need a pound of 4 penny nails," one said, and she reached back for two handfuls and threw them on the balance. "That will be twelve cents."

"Do you have any sturdy wool serge suitable for winter dresses?" a lady asked. Inga went straightaway to pull the exact bolt off the shelf. The customer felt the cloth under her fingers and stretched a piece to test its strength. "Eleven and a half yards, please," she decided. "Are you new here? I never met you before."

Inga unrolled the cloth on the long table, measured it out with a yardstick, cut it, folded it and answered over her shoulder as she walked up to the cash drawer. "Yes, I am new here. I usually make pies with Mrs. Duffy. Eleven and a half times thirty-eight cents a yard will be $4.37 please. Thank you for your business." The customer, with a shocked look, handed Inga five silver dollars and Inga counted back 63 cents into her outstretched hand. Then Inga turned to help a farmer who needed a bridle for his horse.

Every Saturday business was booming. Even when the weather became blustery the shop was busy. Sometimes there was no sale, but she helped just the same. A farmer needed his plow repaired and Inga helped him bolt on a new set of handles. Often customers just needed a bolt and washer, or some glue and Inga sold what was needed and helped put it to use. "You ought to keep this one, Mrs. Ronan," a farmer said one day. "She's worth her weight in gold to a shopkeeper."

The week before Christmas Mrs. Kelly arrived home from Cedar Rapids and sat down with her daughter, Mary, to find out how the business had survived her absence. "I see you've reorganized quite a lot while I was gone," she told Mary. "I should have known a trained teacher like yourself would set the place to rights."

Mary smiled a little as she said "Actually, we've had extra help. Inga, an orphan girl, has been adopted by the Duffys. In your absence she's worked Saturday afternoons to pay back for two of Katie's old dresses and a coat. Tomorrow is her last day of debt. All the improvements to the tools display were her idea. Duffys hit the jackpot when they got her off the train."

The next day was Saturday, Inga's last day of payback. As she came in and brushed the snow off her shawl, she saw someone new sitting near the cash drawer. Someone older, who looked much like Mary. "Good morning," the woman said. "I am Mrs. Kelly. This is my store. I hear from my daughter that you have been a valuable help to her."

"I try hard, yes."

"I consider your debt paid in full a bit early, to give you some free time to prepare for Christmas. Your dresses and coat are all paid for, being used and all. Yet, you have done more than we ever expected. Come here and hold out your hand. Close your eyes."

Inga did so, not knowing what to expect. Surely

she would not be struck with a ruler as sometimes happened in school. She blinked but held her hand steady. Mrs. Kelly reached into the cash drawer and took out a silver dollar. She put it into Inga's hand and closed her fingers over it. Inga had never had so much money of her own in her life. A whole dollar! It would buy a new dress all her own, or a pliers. She wanted to hug Mrs. Kelly, but hesitated. "Thank you, thank you, Mrs. Kelly," she babbled. "I only pay back what I owe."

"You gave full value and much more, Inga. If you need another job I will give you a glowing recommendation."

Inga walked home hardly feeling the snow that drifted down the back of her neck. A whole dollar, all to herself. What a day this had been.

Chapter 12
Swedish Christmas

The pie business continued to grow, now that the new helper provided an extra pair of hands. School was put off on busy days and sometimes all week. Now, after the Thanksgiving rush, people were ordering four or five pies each to get ready for Christmas.

The second Saturday in December Mr. Duffy gave in to Mrs. Duffy's nagging. He went out to the little woods near the river to cut a Christmas tree. It was a scraggly thing missing most of the branches on one side, no taller than Inga. They propped it in a bucket in the corner. While pies were baking, Inga and Mrs. Duffy threaded garlands of popcorn and slices of dried apple for decoration. "I wish we had little candles," Inga told Mrs. Duffy. "In Sweden it is very dark at Christmas. We had many candles. They made us feel warm."

"We have the stove, Inga. Candles on that dry

scrub would burn the house down in a minute."

"It was nice. We called it Lucia Day. The oldest girl wears candles in a crown and brings coffee cakes to the family in morning."

"Candles in their hair? Don't the Swedes worry for their children? I never heard of such a thing!"

"It was pretty. I was too small. My cousin carried the cakes and she didn't burn her hair. She walked very slow. It was our best Christmas tradition. Maybe you don't need it in America because it is very light here every day. In Sweden there is no sun at Christmas."

"No sun? What do you mean?"

"The sun does not come out. It is dark all day."

"That is strange. When does the sun come out?"

"It comes out a little bit at noon after Christmas. Then every day a little more. In summer we have no darkness."

"It's a strange place. You're well out of it. We have proper days and nights here, and no hair on fire."

"Maybe my father will send a Christmas letter. I wish for that so much."

"I would think he could bring himself to write, being an educated man. Perhaps he lost the address."

"He went for a long time to school in Sweden. He can write. Maybe he has forgotten me. I am so far away." Inga sniffed a bit and wiped a tear off her face with her floury hand.

"Just write to him again and write to Mrs. Hobson too. She may send it on. You are safe here. Let's think of happy things," Mrs. Duffy said.

They continued to roll out dough and fill pies as they talked. They mixed batch after batch of dough.

"I need more lard, Inga," Mrs. Duffy said. Inga wiped her tired puckered fingers on a rag and went to get the lard. No sooner had she placed her knife in hand to peel another apple, than Mrs. Duffy needed another peck of flour. "Just put it in here and bring some water," she said. There was no stopping for lunch or rest. They mixed the dough, and kneaded and rolled out the pie crusts, one after the other. Soon pies filled the oven, with more at the end of the table waiting to be baked. Inga mopped her forehead with her apron. Even in winter the room was hot. The work pushed all the sad ideas out of her mind.

While Inga peeled basket after basket of apples her hands grew sore and tired and her mind wandered. What if a machine could do some of this? Maybe a machine could peel the apples and save her hands. Just crank them around with a knife somehow. She could almost picture it—a machine for peeling. What a fine thing that would be!

"Pay attention, Inga," Mrs. Duffy reminded her. "Those pies in the oven smell like they're scorching. You have to move them around. Don't sit there daydreaming."

"Thinking is good," Inga muttered to herself. "Do some yourself." The last bit slipped out, but Mrs. Duffy was too busy to notice. Inga kept on dreaming as she shifted the hot pies from side to side. The wringer didn't work for the dough, but maybe a machine could peel apples. Someone like her papa could figure it out. After all, he was helping figure out train engines. Pies should be easy. Easy enough for a smart girl like me as well, she realized.

The scorched smell of burnt apple interrupted

her thoughts and she scurried to pull out a pie that had bubbled over. Burned apple and melted sugar dripped down inside the oven. Mrs. Duffy gave a deep sigh when she saw it. "There goes hard-earned money, and a lot of work. We can't keep eating up our mistakes."

"I'll peel some more very fast, Mrs. Duffy. I am sorry for so much thinking," She paused. "Do you have an extra knife, Mrs. Duffy? One that I can keep? I want to make a machine to help you."

"A machine, is it? Surely a machine cannot make pies."

"I am thinking of a better way to peel the apples. I just need a knife. Maybe I can get some other things from Mr. Duffy. I have to do a lot of thinking first."

After the mess was cleaned up and the burnt pie was scraped off the oven sides, Inga got out a piece of paper and began to sketch her ideas. Something to hold the knife. Something to hold the apple. Something to carry the knife around the apple. It would have to be strong. And how could it move around—that was the sticking point. It would wobble and be hard to control. She thought and thought. She drew and drew. What would her father do to solve the problem? When both sides of the paper were full, she asked Mrs. Duffy for more paper. "Paper is costly, Inga! This isn't a school. Even in school they write on slates to save paper."

"Do you think I could borrow a slate from the school next time I go? Maybe the sister would have an extra one."

Sister did have one and loaned it out on a Friday with a warning not to drop it. "Slate breaks, you know, even though it is stone. Even stones can break."

That night Inga drew and drew, scrubbed and

scrubbed. Suddenly she had an idea. What if the knife stayed in place and the apple turned? That idea gave her a new push and off she went, drawing and erasing.

On Saturday as they finished their last batch for the week, Tom Kelly and Emmet showed up to get their families' pies. "What'cha doin'?" Emmet wanted to know. "Drawing ideas? I like to do that. I have a whole notebook full and I've started a new one for this school year. Miss Kelly said it's what all the scientists use."

He picked up the slate and held it near the window to get a better look. "Looks like you have some sort of machine here," he said. "Do you like machines?"

"I want an easier way to make pies. Machines are good. My father makes machines. He showed me how to take them apart."

"So what is your idea? Looks like you have a lever and a crank to turn a thing that's sort of round. A long flat blade...a knife? Are you trying to peel apples?"

"Yes," said Inga.

"A machine might bruise an apple if it's strong enough to cut. Besides, I don't think a small crank like that would be strong enough to turn it. You need a lot more mechanical advantage."

"My invention is strong," said Inga. "My invention does not bruise apples and the knife stays in one place and it's the apple that moves."

Tom looked over their shoulders. "My mother sells something like that. You screw it onto the table, put the apple on here..." he pointed. "Then you turn this handle and the apple rubs against the knife and zip, off comes the peeling."

"It isn't a new idea?" Inga was shocked. "I didn't see them when I was working there."

"She's out of them, then. My mother orders 'em from a place in Chicago. It's already invented. My sister Mary bought one."

"All great minds think alike," Emmet said solemnly. "You can ask her when you're in the store again. She'll show you."

"I'll ask Mr. Duffy this very day," Mrs. Duffy exclaimed, listening from across the table. "How much do they cost, Tom?"

"Almost a dollar, I think."

Mrs. Duffy's face fell. "Mr. Duffy will never..." She shook her head sadly.

"I think mother would take pies," Tom exclaimed. "I'll ask her."

"I can work an extra day before Christmas, Mrs. Duffy," Inga said. "It was my idea and it will help my hands."

"What a great idea, Tom, and thank you for your offer Inga. It was your idea that got us started. Six pies should cover the price, and your extra work.... Yes we can do it, and now you'll have more time to go to school."

Chapter 13

The Apple Peeler

Tom and Emmet were taking turns carrying a box toward Duffy's shop when they ran into Mr. Duffy himself sweeping the snow off the porch. "What have you there, boys? A present?" he chuckled.

"It's the apple peeling machine your wife ordered," Tom said. "Ma asked us to deliver it."

"A machine for peeling apples? That wife of mine gets the gol-dangdest ideas!" He propped the broom against the wall and stomped into the house. "Lavinia," he bellowed, following Tom and Emmet into the house. "What do you mean by buying useless machinery without telling me? And where did you get money for it?"

"It didn't cost me a cent, Henry."

"Don't try to trick me, woman. Machines cost money. The boy here says it's to peel apples, of all things. What do you think we got this girl for? She's cost us a pretty penny already for fancy clothes and food. Now you're spending more money to do her work. I won't have it, I say!"

"Henry, breath deeply. Now sit down and we can talk. You'll frighten the children, and they'll tell their parents and what will your customers think? We can at least look at it." He gave a deep sigh, pulled up a stool and plopped himself down.

They unpacked the machine from its wooden box, dumping the sawdust packing into a bucket near the stove for kindling. The peeler was a remarkable object, black and made of iron. They set it at the corner of the work table and Mr. Duffy reached into his overall pocket for a nail to fasten it down.

Mrs. Duffy stepped between him and the peeler. "Don't you dare pound nails into my fine table, Henry Duffy! I hauled that bird's-eye maple table all the way from my aunt in Ohio and I'm not about to have any nailing done."

Inga pulled out a clamp out from under the sawdust. She held it up. "See, this fits at the bottom part. A little hook goes into the hole. It doesn't need a nail."

The adults looked at the usually-silent girl holding up a strong black iron clamp and unscrewing the handle. "Well, I never..." Mr. Duffy blustered. "But it's a smart piece of work," he admitted after another look. "The gears look fine, and they even thought of the clamp. Let me set it straight and you can fasten it, Inga."

Once it was clamped firmly to the table, Inga pushed an apple onto the sharpened rod. A large gear wheel turned a smaller gear. The smaller gear turned the apple against a spring-mounted cutter. When the handle turned, the spinning apple turned against the cutter, taking off a long spiral piece of the peel.

"Well, I'll be jiggered," Mr. Duffy exclaimed, taking the first turn. "It actually works. Somebody was pretty clever to think this up. Probably made a mint of money off it."

"I think so," Emmet interrupted eagerly. "See, here on the side it says 'Penn H. Company' and 'Pat'd.' Every time they make one, they have to pay the Penn H. Company. It's a law. I read about it."

"No surprise on that," Tom said.

"And you say it didn't cost you anything, Lavinia. Who do you think paid the patent?"

"I have no idea, but I did not. I traded pies to Mrs. Kelly and Inga worked extra at the store. It's our own labor."

"Well, you've got the better of me this time, Lavinia. I didn't think you had it in you. A useful thing practically free. Never thought I'd see that." He shook his head, buttoned up his jacket and trudged out to his shop.

Inga tapped her fingers on the table next to the peeler and twiddled the end of her braid where it crept out of the pins. "Maybe I could do this patent work. It is only thinking, and then you make something. Then when I grow up I can buy new clothes with my own money."

"Clothes, clothes, that's what girls always think about," Tom teased. "I'd get a horse of my own and a fast wagon."

"I'll help you get a patent, Inga. I can find out anything you need to know. It would be fun," said Emmet.

"Me too," Tom said. "I even help with explosions. Just ask Emmet."

Inga looked over at Emmet and raised one eyebrow. "Do you want to help me, really? Build my ideas?"

"If they're good, I will," Emmet stated firmly. "Just not ideas about making dresses. I'm not interested in dresses. They're girl stuff. Something with science parts—that's what I like."

"Children," Mrs. Duffy said, interrupting their daydreams, "these pies won't wait. You'll have to either help me or go. There's work here and no machine to do it." She smoothed a bit of flour out on her board and pulled out her big bowl. "A peck of flour, please, Inga—and some lard and water. Good-bye, boys. Give your parents my regards."

"We will, Mrs. Duffy," they said as they backed toward the door. "Can we watch when you peel the apples? Monday after school?"

"Yes, I suppose so. Now be off with ye. I've lost enough time today."

That night Inga wrote another letter. Maybe her papa would get it before Christmas. Mr. Roche at the post office said it could get to New York in two weeks.

Dearest Papa,

Why do you not write? It is almost Christmas and I want to know you think of me. We have a very poor Christmas here without the Christmas songs and the Lucia celebration. I miss the songs you play on the violin.

I told Mrs. Duffy about Lucia. She thinks Swedes let their children burn their hair. I miss the fish for holidays and the ginger cookies mother used to make.

They don't put candles on their tree! It is such a sad tree.

Mr. and Mrs. Duffy are nice, but I would rather have you. Please write to me and say that you didn't forget me.

Your loving daughter,
Inga

Chapter 14
Cyril Falls

Soon the holiday was over, and the bedraggled tree had been burnt up little by little in the stove. A great pounding rattled the door. "I've come for my pies," shouted Brady. "Anybody home here?"

Mrs. Duffy opened the door, and a huge puff of warm steam hit him in the face. "Come in, then, and get a bit of warm," she said. "It may be a January thaw, but wind makes my bones ache." She closed the door firmly behind him.

"It'll have to be a quick warm-up, then," Brady said. "I left Cyril out in the wagon and he'll wear me out complaining if he has to wait." Just then they heard a thump outside, followed by a shout of "Dang it all!"

Inga ran to open the door. On the wet ice next to the wagon, Cyril lay sprawled with one leg askew. A big bump on his forehead was growing by the moment.

He sniffed and rubbed his face with his sleeve. Inga held out her hand to steady him as he struggled to stand up on one leg. He straightened his other leg. Handing him his crutches, she helped him limp into the house.

Brady turned as they came in. "Dang useless good-for-nothing! I told you to wait in the wagon. But you can't take orders, can you. You've gotta get down on the ice and make a dang fool of yourself. Blast the day I ever sent you to that school. If you'd been home on the farm you'd have two good legs today and be worth something to me."

Mrs. Duffy stared at Brady, her eyes wide. In a shocked voice she said, "Bring him in, Inga, and give him the stool near the stove. It won't do to have him gettin' a chill after all he's been through. Then pack up the pies for Mr. Brady while he collects himself. He'll soon be wanting to comfort his son, I'm sure." She turned to Brady. "Can't you see the boy is hurt, Cletus Brady? A fall on the ice is no picnic, even with two good legs. And your own flesh and blood, too." She smacked her rolling pin down on a fresh lump of dough. "For shame."

Cyril dropped the crutches with a thump and sat down on the stool, pushing his feet toward the stove.

"I thought you lost your foot," said Inga. "You have two boots."

"My stars, Inga! How rude. You must excuse her Cyril. The Swedes are so direct. She has no idea..."

"I don't mind, Mrs. Duffy," Cyril said. "I'm used to being a cripple and explaining it and all...Everybody asks me. I've got a goose egg on my head from hitting the ice though. Do you have a cold cloth for it?" Inga fetched a clean cloth, wetted it under the sink pump, wrung it out and put it gently on the bruise that was rising rapidly on Cyril's forehead.

Mr. Brady was still sputtering, but Mrs. Duffy gave him such an angry look that he stopped, embarrassed.

Cyril pulled up his pants leg a bit to show the leg to Inga. "Are you sure you want to see it? It ain't pretty. Here, see. This foot isn't real. It has a shoe like the other one, but the leg part is wood. You can knock on it." He took Inga's hand and pulled it down to knock on his leg. It sounded just like the door.

"Does it hurt?"

"Can't feel anything down there. It's wood. All the hurt is up under my knee where Doc cut it off. It isn't bad any more. It happened a year ago in the storm. But it aches in the cold."

"Does it work like a real leg?"

"It holds me up."

"What about walking?"

"I manage. That's all."

Mr. Brady finished loading up his pies in a crate. He carried the heavy crate out and stowed it in the wagon with his other supplies, spreading a blanket over it. He came back inside. "Let's cut out all this chatter and get on home, Cyril. The girl has no need of a story about your foolish accident. She's just tryin' to be polite." He handed Cyril his crutches and pulled him to his feet. "We'll be off now, Lavinia. I'll thank you to mind your own business when it comes to interferin' with my son."

Mrs. Duffy inhaled abruptly. "Well I never..."

"Chalk the pies up to our credit. I want to get home before the mud freezes up on the roads. The fog is settling in over the fields. I hope it doesn't break up like last year."

"Praise be! I hope we never have another killer storm like that. And to think Emmet Roche saved the whole class. Who would have thought that little weasel..."

"Yes, yes. Little weasel. Just as crafty. That's the word for him. We have to go, Lavinia." He slammed the door after them, handed Cyril up into the wagon and drove off, the horses' harnesses jangling and echoing over the snow.

"That leg isn't good, Mrs. Duffy. It doesn't work back and forth," Inga said once they were gone.

"What do you mean?"

"Down here." Inga pointed to her ankle. "This must move. Does his leg move?"

"Of course it is stiff," Mrs. Duffy explained. "It's wood. Wood is stiff. He just has to put up with it."

"There should be a better way, though," said Inga. "All the soldiers from the war—are they all stiff too?"

"I imagine so. It's sad but we can't carry their burdens. We have our own—eight more pies to make before our day is done." She reached down into the huge sack of flour and scooped up a peck measure full. "Now get me some lard and water and we'll get our day's share finished. We need stovewood too."

Inga complied, but kept thinking. How could a wooden leg with a shoe on it move like a real leg? What do legs really do when they walk? She had never considered that before and stood on one of hers while she flexed her ankle. Then she bent her knee up and down. The lard hit the floor with a soft thump.

"What's the matter with you? Pay attention. Now there's a pound of lard wasted."

"We can just brush it off. The floor here is clean." She moved her ankles some more, up and down, side to side.

"Inga! Stand still and do your job. You can think about legs after the pies are done!"

Inga nodded and went back to the barrel for more lard. She stirred up the fire in the big stove. "Some way to make it move..." she muttered to herself as she headed out for an armful of oak from the woodpile.

Chapter 15
Flicka

Inga picked up the last load of stovewood, tucking the ends of her shawl under the small logs to keep it from blowing off. A skinny calico cat streaked by, followed by Tom. It dashed under the front step of the Duffy house and hid, shivering, in the farthest corner. It was the same cat Inga saw Tom's brother running after all the time. "Why do you chase that cat, Tom?" Inga called.

"For fun. It's just an old barn cat anyway. I never do anything to it."

"It may be cold or hungry. If it can live here, it will keep mice away from the flour."

"You don't want it. It's probably loaded with fleas."

Inga sat on the step and let her hand dangle down at her side. She wiggled her fingers and made soft little Swedish sounds.

"What are you doing, talking Swedish to an Iowa cat?"

"Just wait and be quiet. Maybe she'll come out." Inga waited, and wiggled her fingers, and waited. Tom fidgeted as he watched, not expecting anything to happen, but curious.

Mrs. Duffy stuck her head out the door to see what had happened to Inga. "You must be cold, and there's work waiting here. Oh, hello Tom. I don't have anything for your mother today. You'd best be getting on home."

"Why are you sitting on the step, Inga? You'll take a chill."

"I am getting cat. We need it to catch mice in loft."

"Well, don't be too long about it."

As Tom walked away, Mrs. Duffy went back in and soon Inga felt a soft tickle on her fingers. "Come out little cat," she said softly in Swedish. "I won't chase you. You can live in my place and have all the mice you want." Slowly the cat came out and rubbed itself against Inga's cold leg. It started to purr and let Inga pet it. It was all bones beneath the scruffy fur. Its ear was notched and its fur was rubbed off in patches. "Will you be my cat?" she asked in Swedish. "I can talk to you and no one will know what I say. You can be my secret keeper." She looked straight into its green eyes. "Your name will be Flicka and I'll keep you safe."

The door opened a crack and Mrs. Duffy looked out. "You're going to freeze out there. What are you doing with that cat?"

"She is mine, Mrs. Duffy. I hire her to keep mice out of flour. She won't eat much."

"I can see that. I see her bones from here. Bring her in then and we'll give her some milk. After that she will have to hunt for herself."

Inga brought in her load of wood and then went back outside. Flicka was nowhere to be seen. "Come in, little Flicka," Inga said in Swedish, using her softest voice. "I won't hurt you. I do not chase cats. I have lovely fat mice for you to hunt in my house. Come into my house, little cat."

She heard a soft rustle under the step. She kept talking softly, holding her shawl tight against the chill. In a few minutes she felt a soft tickle against her cold leg and heard a little purr. She kept talking while slowly letting one hand hang down at her side. It was the hand she used to bring the lard to Mrs. Duffy. Flicka gave her fingers a tentative lick. Inga didn't move, but kept talking softly in Swedish. Flicka licked her fingers some more and rubbed against her leg. The purr got louder.

Inga reached down and slowly rubbed the cat's white, orange and black fur down her back from her neck to the top of her tail. The cat stayed. Then Inga picked up the cat, not minding the fleas. She gathered Flicka in her arms and went inside, shutting the door behind her.

Inga filled a saucer with milk for Flicka, who lapped it up eagerly.

"I see you've tamed the shaggy thing," Mrs. Duffy said. "But I won't have fleas in my house and you won't want them up in the loft with you either. You'll have to get her clean."

Inga got a rag, dipped it in the warm soapy wash water in the sink, and started to scrub Flicka gently. Little by little, from the top of her soft ears to the end of her scraggly tail, Inga scrubbed the cat. It took some coaxing and little bits of cheese, but finally Inga had Flicka clean and damp. "She'll dry off in the loft, I think," she told Mrs. Duffy. The cat followed Inga and her cheese bits to the ladder and Inga carried her up into the loft. Flicka immediately began stalking around the flour sacks. "Just get on with your work, Flicka" Inga told her. "I have work to do myself." Inga scrambled down from the loft to help take out the pies and put in more. "I think I have a friend now," she told Mrs. Duffy.

"I thought you were friends with Emmet and Tom—maybe Katie Kelly at the store. A cat is not a friend."

"She will be the best friend. I can tell her my heart and talk Swedish to her. She won't make fun of me. This is my happiest day, Mrs. Duffy. My very happiest."

"Are you not happy to have your home with us, Inga? We have done all our best for you, even sending you to school."

"Yes you are good but I have a sad space in my heart. When I hold Flicka she purrs and I feel happy inside."

Chapter 16
The Apple Peeler in Motion

On Saturday morning early, Tom and Emmet stopped by to pick up their pie orders. They found Inga hard at work peeling apples with the new peeler machine. Flicka took one look at Tom and dashed toward the loft. "I guess you have a real scaredy-cat," he joked to Inga.

"You make her to be scared, Tom. She isn't scared of me. She isn't even scared of Mr. Duffy."

"Can we watch the peeler machine, Inga?" said Emmet.

"Only if you promise not to scare my cat."

"Doesn't it get your fingers?"

"I am very quick."

"How fast is it?"

"Faster than a knife. Watch."

"Efficiency is important," Emmet commented. "If it isn't more efficient than fingers, then it's no good."

Inga grabbed an apple and stuck it firmly on the prong. She turned the crank four solid turns and the peel streamed off the apple in long ribbons. In less than a minute she had a clean peeled apple ready to cut up for pie.

"That's a wonder," Tom said. "It must be fun. Mother never uses hers. She's too busy tending store to make pies."

"Oh yes. It's a lot of fun," Inga said with a crafty grin. "I peeled a whole bushel yesterday."

"Would you let us try it?" asked Tom.

"I'd like to see how the gears work," Emmet said.

"Or we could have a race and see who peels fastest," added Tom. "Please? Just ask Mrs. Duffy. Pretty please?"

Mrs. Duffy was up to her elbows in dishwater, preparing pans for the next batch. Inga went over to ask and there was whispering and bobbing of heads.

"She says you may try it but you must pay."

"Pay? For peeling apples?"

"For using her new machine. You have to give me a penny for three apples."

"That's robbery!" said Emmet. "It's an outrage! We can peel them for free with a knife."

"Yes, you can. If you want to... at your own house. The machine cost money." Inga stood firm, hiding the machine with her back. "You can decide."

Tom and Emmet went over near the door to have a discussion. "I don't have any money," Tom admitted.

"Me neither," said Emmet. "Maybe she'll trade something."

"They went back. "How about a good marble for every five?" Tom offered.

"I don't play marbles," said Inga.

The boys had another conference while Mrs. Duffy dried the pans she had washed. "You'll have to get busy or go home, boys," she said. "This isn't a circus sideshow."

Just then Tom had a brilliant idea. "We can pay her with lemon drops from our store, Emmet. Mother never counts them. She buys them in a sack. Mary won't notice if I sneak a few out in my pocket."

He went back to Inga. "How about a lemon drop for three apples? Emmet will count. He can count anything."

Mrs. Duffy, listening, hid her smile. "Yes," said Inga. "Lemon drops first." The boys ran off to get lemon drops while Inga and Mrs. Duffy sliced the peeled apples and filled the pies.

"You have a sharp wit, Inga," Mrs. Duffy remarked. "You should have a business when you grow up."

"Maybe," said Inga. "I would like to make useful things. Then people will pay me for them."

The boys stormed in. Tom laid out the lemon drops on the table. There were 16. One was covered with gray pocket lint. "Take this one back, Tom," Inga said. "I don't want dirty candy." That made fifteen to trade. Tom popped the dusty one in his mouth.

"Forty-five apples for us anyway," Emmet crowed. "I wonder how many pies that will make. How many, Mrs. Duffy?"

"About six or seven in a pie, Emmet. You figure it out."

Tom had his fingers out and was ciphering as fast as he could. "It won't be even," he complained. There will be apples left over no matter what."

"We can just eat them," Emmet suggested. "It's only three leftovers. Seven times six is forty-two and six times seven is forty-two, so there are always three left over."

"I can't make money if you're eating up all the apples, children. Just peel the forty-five and we'll stuff in the extra pieces someplace." She shook her head as she watched them count the apples and start peeling. Will wonders never cease, she thought, watching them counting and multiplying and dividing, moving the apples and lemon drops into piles to make sure the count was correct. Who would have thought an apple peeler would become a lesson? No telling what changes all these machines will bring!

As they worked away, Inga said, "I am thinking about that boy Carl. How did he lose his foot?"

"Who's Carl?" the boys asked in chorus. "You mean Cyril? Oh that's a long story. It was around this time last winter, when we had the blizzard."

"A blizzard?"

"Snow and very cold. Too much snow."

"Like Sweden?"

"Worse than Sweden. Some people died here. We had to sleep in the school."

"But how did Cyril lose his foot?"

"He tried to go home from school. He didn't listen to Miss Kelly."

Tom joined in. "He never listened to our teacher. He fell off his horse in the storm and nobody knew he was out in the blizzard. When Jim Ronan found him next morning, his foot was frozen."

"It was stiff and blue," Emmet added. "Like Napoleon's soldiers."

"Who is Napoleon? You had soldiers here?"

"Never mind Emmet, Inga," said Tom. "He has too many facts. No soldiers. No Napoleon. His frozen foot had to be cut off. Doc did it, like in the war. Now Cyril has a wood leg with a wood foot at the bottom."

"I saw it. He made me knock on it. It sounded like the door. That makes him useless, Mr. Duffy said. He cannot do anything. It is sad." She filled her last pie with peeled apples number 44 and 45, and wiped her hands on her apron. "Maybe he can get a better foot."

"They're all made of wood," said Emmet. "Can't help it."

"We must think. Maybe a better one can be made."

"That would be some pumpkins," Emmet said. "We can make good wood legs and make a heap of money for them. All the armies will buy them for the soldiers. When do we start?"

"Maybe I will let you help," Inga said. "I will think of the idea and you can help make it. My name will be on the side like Mr. Penn H. Company says on the apple peeler."

"Girls can't invent things," Emmet said. "They never invent things. They can help, though."

Inga straightened up and looked severely at Emmet. "Girls can invent. Boys can help. It is so in Sweden. I will invent a good leg. You can help me if you want to. I will let you help for free. You don't have to pay to work on my leg idea. But my name goes on the side, like Mr. Penn H. Company."

Mrs. Duffy dusted the flour off her hands. "It's best if all you inventors finish for the day," she said. "You must have homework to do, and parents who need your help with chores. You can think of all this on your own time. Inga and I have a supper to cook and then I promised to help her repair her quilt. Now be off with you. Give my greetings to your parents." She shooed the boys out the door, letting Flicka sneak out behind them for a little fresh air. To think I never had chick nor child and now I have three under foot most days, she thought to herself. I can see why my friend Kate Kelly always looks so tired. She hung three damp dish towels on the backs of chairs to dry near the stove and went in search of her sewing basket.

Chapter 17
Ship from Sweden

Inga took her baby quilt from under her pillow and brought it down from the loft. "Here, Mrs. Duffy. This is my best quilt. My mother made it when I was a child. Flicka has played with it. The insides are coming out."

"That's a bad cat you brought into this house, Inga. She's wild. She doesn't have any cat manners. Yesterday she brought in a dead bird. She won't last long in this home I fear."

"No, she's a good cat. She had a bad life, living outside. Tom and his brother chased her. She just needs somebody to love her."

"Well, that's as may be. Here, let me see the damage."

The room was darkening quickly as the winter sun went down. Mrs. Duffy lit a kerosene lamp and set it on the work table. She held the quilt up to peer at it closely near the lamp. A bit of stuffing worked its way out as she handled it, and floated to the floor. She peered at it, then went to find her spectacles for a better look. Flicka pounced on the stuffing, scattering it under the stove.

"Did you always have spectacles?" Inga wondered. "You do not wear them."

"They just help me read but I don't read anymore. I have too much work to do. I used to love books so." She sighed and raised the quilt up to get a better look, blinking slightly. "Two big rips, from claws I'd imagine. Clean cuts straight across the backing. It's lucky she didn't get the top side. Someone put a lot of work into this, making a pattern out of all those bright little cloth scraps."

"My mother had very fast hands. It is very sad if I think of her."

Mrs. Duffy dug a piece of flour sacking cloth out of her workbasket and cut two pieces bigger than the rips. "Here, we can hold the quilt between us and each work on one end. Lay the cloth over the rip, turn in the edges and sew it down. Leave a little opening to add more stuffing at the end. We can talk while we work. Come closer to the stove."

She shoved her rocking chair near the stove while Inga dragged a stool close to her. They put the quilt between them and Mrs. Duffy started to explain what to do. "I know how to do this," Inga said. "My mother taught me. Remember the dress from Katie?" She continued, "In Sweden girls learn to sew and knit at school too. Boys learn to pound nails. I liked pounding better, but I can do this." She threaded her needle and began to stitch.

"You were telling me of your mother," Mrs. Duffy reminded her, adjusting her spectacles on her nose. "And your life in Sweden when you were little…"

"Mor, I mean mother, was good and funny. She made cloth with threads on a big wooden loom. She showed me how to do everything, even cook. My father fixed machines that were broken."

"So why come here to America?"

"Father's shop closed. There were no more jobs. It was very hard to live. No family could help us. Papa read in the newspaper that America needed mechanics and engineers to fix machines in factories. So we put all of our clothes in a big trunk and bought a ticket. We sailed from Goteborg, on the west side. It is full of ships."

"You must have come to New York. That's where they all come," said Mrs. Duffy, nodding. "I had cousins who came the same way. It is the usual thing."

Inga stopped talking and held her sewing to hide her face. A couple of tears dripped on the quilt. "It was not happy," she said. "Only Papa and I made it to New York."

"What happened to your mother? It may help to talk about it."

"On the ship many people became sick. I know they get sick from the moving ship, but mother had a fever. Every day we put a wet cloth on her head to cool her, but one day she didn't wake up at all. The sailors came down into the steerage, wrapped her in a canvas bag and took her up on the deck. We all stood still and said a prayer and they dropped her into the ocean. They said it was ship fever and they hoped it wouldn't spread. Three other people died after that. Then we got to New York and came into America." There was an uncomfortable silence while Mrs. Duffy thought of what to say next. She could only imagine what it must have been for Inga to see her mother pitched into the ocean.

"I lost my little Bridget," said Mrs. Duffy softly. "She was so sweet and small. Doc told me her chances were slim, with her coming too soon, but we had our hopes up. We thought our hearts would break when we put her under her tiny stone over in the churchyard. Everyone said 'Ye'll have another' and 'You're young yet, you'll have a houseful' but no more came, until we found you at the train."

Inga hung her head and stabbed the needle over and over into the quilt. "That is so sad."

They looked at each other with teary eyes. Inga sniffed and Mrs. Duffy wiped her eyes on her handkerchief and stuffed it back in her sleeve. She polished her spectacles on her apron and straightened her back. " 'Tis all past now. Nothing to be done. We must carry on. Now tell me, you stayed with your father then..."

"He found a place for us to live. Mrs. Lundgren was nice. She cooked for everyone."

"Everyone?"

"All the men. Each one had a room, but some were two or three together. Papa and I had a room for ourselves."

"I should hope so. A child among all those strangers!" Mrs. Duffy stole a look at Inga, who kept stubbornly putting stitch after stitch into the quilt her mother had made. Inga looked up at her.

"Mrs. Lundgren help me learn English. She

taught me to cook American things, like pie and pork chops. Sometimes I went to the American school."

"Ah, now I remember her. You told me once that she was fond of *schnapps*. Then your father sent you off on the train. I didn't know about your mother when you first came here. Why didn't you tell me?"

"Telling doesn't change anything. She is gone, into the ocean, and I am here. Papa told me to be brave. He said I am strong and smart. Some day he may come for me himself. Maybe he has forgotten me. I wrote two letters. He didn't write back. I think I will have to make my own way." She jammed her needle into the pincushion and brushed the thread scraps away. They finished the repair, stuffing wool scraps under the final stitches. They shook out the quilt.

"I am thinking about that boy Carl- no, Cyril. His father calls him 'useless' and so does Mr. Duffy. That is a bad word. In Sweden it is the worst. I am sad for Cyril. Can't he get a better leg?"

"I think he has a good one. Doc Evans ordered it from Chicago. They make a lot of them there. The soldiers are old now, but you know we had a big war here and so many lost their legs. False legs are a big business."

"I will think about this. Maybe I can make something better. Emmet and Tom always like to help. But first I must think. Maybe if the foot could move..."

Chapter 18

The Grand Experiment

"Emmet, let's fix Cyril's foot."

"You mean make an ankle?"

"Yes, let's figure out how an ankle works."

"Then we have to start by observing a real one." Emmet wiggled his foot. "Watch me walk." He walked in slow motion to the tree at the edge of the playground, pointing his toe like a dancer and taking big steps. Inga copied him, trying to feel how her leg worked. They walked back, thinking about every move, right into a puddle they hadn't noticed. They both stamped the wet mud off their feet.

"What are you doin'?" Tom interrupted. "Plannin' a square dance out here?"

Students began to gather around, curious, giggling. "How about a reel?" a boy shouted. "I'll whistle a tune for ye."

Emmet froze in his tracks. "We can't do

science out here in public, Inga. People don't know an experiment when they see one." He turned to the kids and spoke sharply. "We're not putting on a show here. It's an experiment." A spitball hit him above the left ear as the kids drifted away. Some comments drifted back about "crazy Emmet and the hag."

Inga hung her head. "They don't like me," she said. "They say I am hag. My name is Hagstrom. I don't know why they say my name in a mean way."

"They just don't know anything," Emmet said. "A hag is a skinny old lady who may be a witch. You are too young to be a hag. They can't even call accurate names."

"I think I can make a better kind of foot."

"I thought you were planning to make a pie machine," said Emmet.

"You said that at Duffy's," Tom remembered. "I didn't think you meant it. Nobody can design feet. That's God's job."

"Sometimes we need to help. It would be a great business. Better than pies."

Emmet nodded. "If we could do a business like that, it would..." The bell rang just then, so Emmet and Tom hurried to line up at the boys' door. Inga followed the girls through the other one, minding how her ankle moved.

All afternoon Inga's mind wandered to the problem of how a false ankle would work. The ankle bends, that's easy. But a real ankle moves the foot forward and backward. A false ankle would need a spring or lever to pull it up and down every time. Moving it round and round was just too hard to think about.

When a person steps, the foot pushes on the floor

to move forward. How could that work? Some kind of automatic machine is needed to push and then bend back for the next step. Working her penholder up and down like a leg, she tried to picture how that would go but she couldn't get it straight in her head. A couple of times she dropped blobs of ink on her work and got a dirty look from Sister. "Stop your woolgathering, Inga," she said.

"What does 'woolgathering' mean?"

"Thinking one thing and doing something different. Your head and hand have to work together. You can dream at home."

Inga tried to keep her mind on her list of spelling words, but the image of a working wooden leg kept coming into her mind. Once she almost tipped over her ink bottle. I'll have to make a little one and try it, she thought. If I can make a little one work, then a big one should work too.

Finally the bell rang. The children strapped their books together and scrambled for their coats in the cloakroom, but Inga could not stop thinking about ankles. "This takes more than one brain thinking," Inga said to Emmet as they left school. "Do you want to help me think?"

"I'll help you think. But if you make something I have to put my name on the side too. 'Roche and Hagstrom Pat'd.' If they write it small it will fit."

"Hagstrom and Roche."

"Roche and Hagstrom." Inga gave Emmet a dirty look, turned her back, and started to walk away.

"Well all right. It's alphabetical I suppose. What's the problem? I can find out all the facts." He hurried to catch up with her.

"We have to make a little leg that works."

"Like a doll leg? I don't play with dolls. Dolls are stupid. Besides, you don't know enough to make a leg. It needs a lot of research. You hardly know how to read English."

"Then I'll let Tom help. He knows how to do things." She marched off toward Duffy's, commenting over her shoulder as she left, "I can do it myself."

Mrs. Duffy was waving from her door. "Hurry up, Inga. Time's a'wasting. I need some wood for the stove and there's a bushel of apples waiting. Don't stand lollygagging with Emmet."

"Yes'm. I'll get to work." Inga dropped her strap of books inside the door, picked up the long axe and headed back out to the woodpile. She split the pieces of wood the long way, setting them on end one at a time and slamming down the axe hard. Funny, she thought. They only break the long way. If I build a better leg, I'll need to get a saw. Somebody must have one to loan me. She set up the next log.

Whack! The log shifted and axe came down wrong, almost striking her foot. Inga stepped back and took a deep breath. I almost needed a new foot myself, she thought. I have to keep my mind on my job. By the time she chopped an armful of log pieces, she was wet with sweat under her shawl and her braids hung in tatters around her face. Stacking the logs in her arms, she trudged up the steps to the door and knocked with the end of a log. "Mrs. Duffy, open up! I have wood."

The door opened. "Heavens, my girl. To think you split all those with the big axe. I thought we had some small pieces out there. Why didn't you use Mr.

Duffy's saw? That axe could have taken your foot off! What was I thinking telling you to get wood? Here, let's pile them by the stove. You must be exhausted and your hair is all awry. I'll get the coal out of the shed myself. Just set a bit and have a cookie."

A saw, Inga thought. I didn't know he had one, but it's just what I'll need. She reached into the tin for a cookie—sugar cookie with raisin filling. It was still warm and the juicy filling warmed her soul. I am so lucky, she thought. Now if only I can make a useful thing to take care of myself when I grow up.

Chapter 19
A New Foot

"It needs to bend," Emmet announced abruptly to Inga as he walked into school the next day. "From my research I find that ankles must always bend. It was discovered by the early Egyptians."

"Ja, of course. I don't need Egyptians to tell me that," Inga retorted. "I can see it on my own foot. Just look down at your own foot. This is not a new idea."

"Just helping, Inga. I'm happy to be of service." He gave a little bow with a smirk on his face.

I'd like to smack him, thought Inga, if he wasn't wearing spectacles and wasn't three inches shorter. Need help indeed, I'd like to see him split a few logs. But she kept her manners. "Thank you for the research, Emmet. I will think on it."

At recess, Inga stayed inside, pretending to have a headache. In her notebook, she turned to a new page and drew a human leg, marking the places where it needed to bend-- knee, ankle, base of toes. She closed her eyes and moved her foot carefully, trying to feel the bending. She went up to the teacher's book cabinet and looked at the hinges on the door. She measured how wide they were and then, sitting on her bench, measured the width of her own ankle. They were almost the same. Maybe a wooden leg from Chicago was wider than her own ankle, she thought. She recorded her measurements in her notebook and closed it just as the children came trooping back inside. I must think how to measure Carl's—no, Cyril's—wooden ankle.

An idea was forming in her mind. An idea so bold and dangerous that she couldn't tell anyone. She could cut off Cyril's wooden foot, screw in a hinge at the ankle between the foot and the leg, and then he could walk better. That would be an invention! She could make papers for a patent and send them in. Factories would make wonderful wooden legs with Hagstrom Pat'd on the side, or maybe under the foot. She would help hundreds of soldiers and she'd get very rich. People would look up to her. The thought gave her goose bumps.

"Are you getting a chill, Inga?" Sister asked as she passed. "Perhaps you should go home to Duffy's and ask Mrs. Duffy to make you a poultice. We don't want a fever going through the school."

"I am not sick, Sister. I am only thinking." Sister gave her a quizzical look. Inga ducked her head, pulled out her McGuffey's and started tracing the words with her finger. Sister just shook her head and turned away. "Order, please, children," she announced. "Take out your slates for arithmetic. We'll continue where we left off, learning how to convert volume of a wagon to bushels."

Inga waited after school for Emmet to produce his research but he was late. So when she went to the woodpile for stovewood, she searched for a piece about the size of her lower leg. Then she found a shorter one for the foot. I can test it first, she thought. I don't need Cyril until I have a good working one and I can try this myself. She lined up the wood beside her leg and put the short piece in the place of the foot.

Emmet arrived grinning. "I looked in at Doc Evans' window as I was coming here. Legs aren't like what you have. You're supposed to have two leg bones, not just one big chunk like a log. One bone in the top of the leg and two bones in the bottom. Lots and lots in the ankle. That wood you have just isn't right."

"I am thinking in my own way, Emmet, thank you, and you're late. I like to have my hands on my ideas. I can think better that way." She worked the leg part back and forth against the short piece as she talked.

"You need drawings and measurements," Emmet said. "Just putting logs together isn't scientific. You need proper drawings with measurements and arrows."

"I need that wood for the stove," Mrs. Duffy called from inside. "Quit your dawdling!"

Inga gathered her load of wood and handed the leg and foot pieces to Emmet, saying, "Here, be useful."

Emmet followed her inside, absently patting Flicka as he passed. "Do you have any spare pie samples, Mrs. Duffy? Any that got a bit burned or came out crooked? I'll get them out of your way, help to clear some space on your work table."

"Your dear mother, rest her soul, would be ashamed to hear you beg like that, Emmet! You must wait to be invited. People will think you were brought up in a barn. Help Inga stack that wood near the stove and get me some coal from the shack. You can earn your pie." She went back to work as the children stacked the wood.

"She doesn't like me much," Emmet told Inga.

"You have to work hard, Emmet, not just think. People are tired of so much thinking and talking. It fills up their heads." She picked up the two pieces of wood that matched her leg and foot and held them against her own leg. "I'll need some tools. Does your father have a chisel, Emmet?"

"No."

"Maybe Mr. Duffy has one."

Mrs. Duffy, overhearing, said, "I wouldn't bother him, dear. He doesn't work in wood you know. You might ask Jim Ronan next time he brings his horse in. He's a grand woodworker and a great hand to loan his tools. I fear Mr. Duffy would take a dim view of any chisel work. It's best to keep that project under your hat."

"Under my hat? What does that mean?"

"Oh, it's just a way of saying," Emmet explained. "Keep your idea in your own mind and don't let others know it. You know your mind is in your head, don't you?"

"I don't understand American ideas, Emmet. Some are very strange." She shook her head as she set the two wood pieces on a high shelf across the room and went back to stacking the stovewood with Emmet. "How will you help for this leg idea, Emmet? I don't need talking. I need a wood holder."

Emmet stayed quiet a moment. Nobody had ever told him they didn't want his talking. Well, maybe Pa said it now and then, but then that was just Pa's way. Most people said he was smart. Of course, not everybody. Some people just stared. He thought a little more. "Maybe I could draw what you make and mark down if it works or not. That's what I did with my experiments last year. Draw it, try it, and write what happens. Miss Kelly said that was a good thing to do. I don't know if that would work for a wooden leg. We could remember to not make the same mistake twice."

"Will you keep the pictures in your notebook?"

"Yes."

"I think we must put them into my notebook. Mrs. Duffy got one for me when she got the apple peeler. Remember? I have some ideas started in pictures. Some writing is in Swedish."

"All right. Your invention, your notebook. I'll sign the pictures though, they're mine."

"We need to ask Mr. Ronan for a chisel."

"His farm is out in the country past the stone school. It's a long walk and the roads are still frozen. Ask Mary Kelly, she's his wife."

"I know. But she's not working at the store anymore. Her mother is back. I will give her mother a letter for him."

So they wrote a letter on a page carefully torn out of Inga's notebook.

Dear Mister Ronan,

I have the honor to ask a favor of you. I need to use a chisel to help my invention. It is a new wooden leg for soldiers. When my invention is made, I will give your chisel back.

Tack,
Ingaborg Aleksdotter Hagstrom

"That should do it, Inga. What is a Tack?"

"It is Swedish for thank you. Should I make it in English? Will he understand?"

"Let's leave it in Swedish. We don't have paper to waste on a short word. Ronan is smart. He'll figure it out."

Chapter 20
The Chisel

Jim Ronan stopped in at the blacksmith shop a few days later. "Are the children around, Mrs. Duffy? I have here a letter requesting a chisel. It surprised me. Whatever would a young foreign girl want with a sharp tool of that sort?"

"She plans on making a false leg," Mrs. Duffy said. "She hatched up a plan with Emmet Roche—you know him."

"Ah, yes. I don't put anything past that one. I don't want my good chisel to be a leading item in a story about another child in Floyd needing a wooden leg."

"But he means well..."

"Lord save us from people who 'mean well'. A chisel can cut wood like butter, and a hand or leg even easier. My wife would never forgive me if a child got hurt with a chisel of mine. Perhaps I should help them a bit, keep my own hands on the tools. Will they be coming directly from school today?"

"They'll be here at 3:35 right as rain, with their hands out for a piece of scrap pie. Why don't you set a bit, take the weight off? It's almost 3 already."

Soon the children burst in with Emmet and Tom in the lead, out of breath, and Inga plodding behind, hauling her strap full of books. They crowded around Jim Ronan. "Remember me?" Emmet said. "Remember when you saved us? This is Inga. She needs a chisel to make a wooden leg. Will you loan it to her?"

"Hello, Inga," Ronan said. "I've heard that you can fix things, but a leg is something special. What do you plan to do?"

Inga unstrapped her books and pulled her notebook out of the pile. "I have made a picture of the idea," she said, thumbing through the pages. "I have it in here." She showed Jim Ronan page after page of sketches, starting with what looked like sticks fastened together until the last drawing, which looked remarkably like a real leg. "See, Mr. Ronan? I have tried many ways to understand the leg. I think if we could have a real leg of wood like Cyril's, we can test if my idea works. It is better to test."

"Tell me. What is your idea?"

Emmet cut in. "We have the sizes. My leg from the back of my knee to the floor is 10 ¼ inches. My foot is 6 ¼ inches from the toe to the back. Width doesn't matter."

"We put a hinge where the ankle comes together," Inga finished.

"What say you, Mrs. Duffy? Your husband

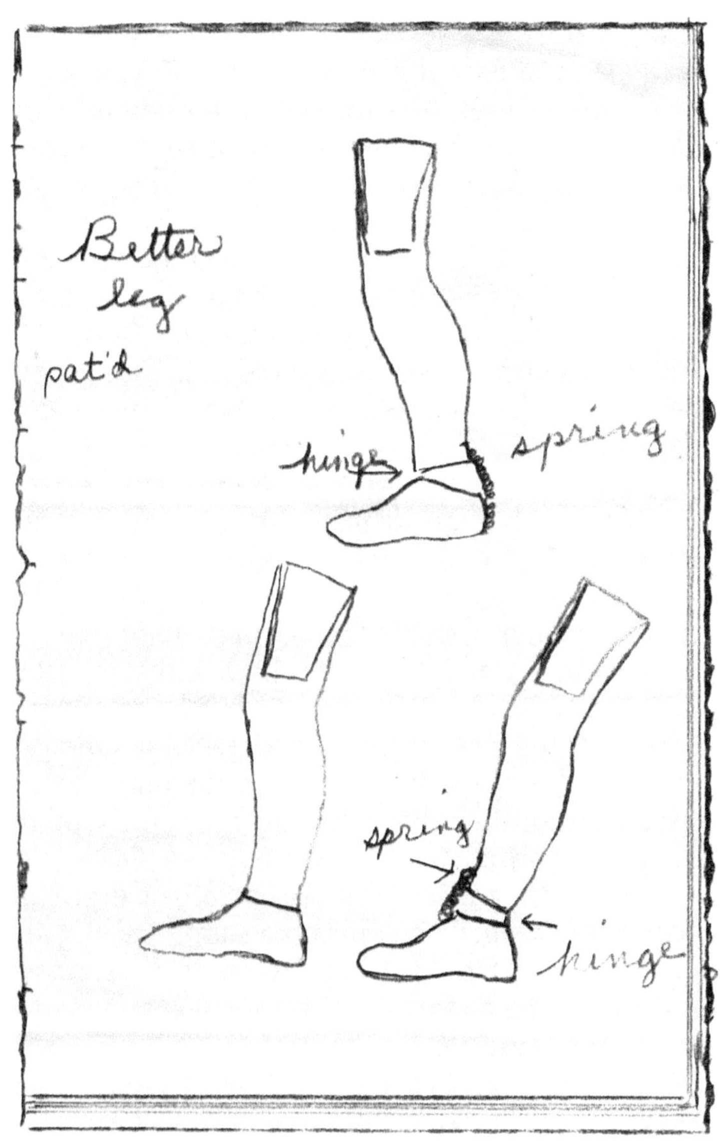
Better
leg
pat'd
hinge
spring
spring
hinge

trusts me," said Jim Ronan turning to Mrs. Duffy.

"Just keep the kids under your watchful eye," she replied.

"Do you have a piece of wood?" Ronan asked. "I'll start the chisel work for you. The wood may slip around. The chisel's mighty sharp and a little mistake can cost you a finger."

"Mr. Duffy has a tool to hold the wood," Inga suggested. "I can do it myself when he is not home."

Mrs. Duffy startled, knocking over a cup of flour. "You must not interfere with Mr. Duffy! His tools are his own. He'll have a conniption fit if anyone uses his shop for another project. Don't even think of it."

"He knows I'm careful," Ronan said. "I won't use anything but his bench vise, and I'm a good customer."

"Ask him yourself, then, Jim. Swear the children won't be touching things in there."

After a while Jim came back into the house smiling. "He has to do an errand but he was happy to show off his new vise. It has a special handle that he designed himself. But I must be the one with the chisel in hand. Wouldn't want to put a scratch on his nice wrought iron."

Once they heard Duffy's horse trot down the street, they trooped back into the shop, and Big Jim Ronan clamped the small log into the black iron vise. He followed the pictures in the notebook and started to carve the short log.

"Does it have to look like a leg?" Tom wanted to know. "It seems an awful lot of work just to test an ankle."

They all looked at each other. Inga thought about it. You could see her brain humming as she considered. "We only have to see what the hinge does," she said finally. "The shape doesn't matter. You are right, Tom. Maybe just two pieces the right size will do. That way we can test quicker."

Ronan put the longer piece in the vise and cut it to size. Then he did the same to the foot piece.

"Now the hard part," Inga said. "We have to put the hinge in just right so the foot piece can go up and down, but not too far. It must go the right direction or Cyril will trip on it."

Inga pointed to the place where the hinge would go, putting a careful line with a pencil just where the bend should be. "Put the hinge here, Mr. Ronan." He screwed the hinge in place. They stood the two pieces of wood on the floor and moved the leg piece back and forth. They compared it to Emmet's leg that was wiggling beside it. The wood "foot" flopped every time it was lifted. "We need a spring at the back," Inga realized. "And one in front too. It moves two ways."

Jim Ronan rummaged in a box of leftover parts that Mr. Duffy had in the shop. Under some brackets and a damaged padlock he found two old floppy springs, one three inches long and one almost six. They put the longer one from the bottom of the foot, over the heel and up the back of the ankle. The shorter one went from the lower leg to the top of the foot. Then the foot model worked nicely.

As they were all peering down at the ankle model, they heard Mr. Duffy unsaddling his horse behind the door. He was talking to someone. It was Brady, Cyril's pa!

"Put it under your apron. Quick, Inga!" Emmet sputtered. Not a moment later Duffy and Brady came clumping in.

"Don't you kids touch my tools," Duffy grumbled as they came into the shop. "They cost me a bundle."

"I'm supervising them," Ronan assured him. "And we're just now finished with my project. Thanks for the loan of your new vise. It's a wonder to behold— so smooth, and the screw holds everything in place. I don't know how you can do such fine work with hot iron. It fairly amazes me. I'll be going along now. Must say goodbye to your missus."

Jim Ronan turned back toward the house door with Inga and the boys close behind him. Duffy was too busy showing off his new vise to Brady to notice the odd shape under Inga's apron.

Once in the kitchen, Jim Ronan asked Inga if she needed a real leg shape carved. "I can do it when it's too wet to plow," he said. "My wife misses the kids and their projects now that she's home with me and my mother."

Inga and Emmet looked at each other. "We are only making a test," Inga said. "Wood pieces will do."

"Let's have a think on this for a week or two. Just among ourselves. Look at all the possibilities. Cyril is in no special hurry. He doesn't even know what you're thinking. If there's anything I can do to help you know I will. I'll be getting along now. It's almost milking time and my Mary will need a helping hand. I'm glad her mother came home. I was getting tired of cooking."

"That's not very authentic," Emmet argued after Jim Ronan had left. "It won't act like a real leg."

"And so? It only must bend in the right place, Emmet. It will never be holding up Cyril."

"It won't? Then why make one?" Tom asked. "I thought you were building a new kind of leg."

"We will cut Cyril's leg and put in a hinge and springs. The rest is already made," Inga explained.

"Cut the leg Doc Evans ordered? Cut the expensive fancy leg from Chicago? Brady will kill us when he finds out," Tom said.

"My pa said Brady paid $58 for it," Emmet added. "He was telling everybody over at Kelly's store."

"He must not know until it is finished. We must make very good plans."

"Then we need a safe place to fix it. Someplace where Brady won't see us," Tom warned.

"I think I have it. We can work in Ronan's barn," Emmet said. "It's closer for him than coming into town. Miss Mary, I mean, Mrs. Ronan, won't mind. She likes scientific projects."

Mrs. Duffy overheard as she was stoking the stove and came over to the group. "You'll be in a pack of trouble with this. Brady is the kind to hit first and talk afterwards. He might even get the sheriff." She shook her head. "I don't recommend it at all. It is far too risky. You know what Brady's made of. And you will have to ask Jim Ronan if you can use his barn first."

Emmet sighed. "I remember Brady was mighty put out when we burned up his hay wagon. He told my pa to whip us. Lucky we got off with just the job of cleaning out his barn."

The rest of the week Emmet and Inga puzzled over how to get the idea to Cyril. Finally Emmet asked his father if he could ride out with him to the country school to give a book to Cyril. "Sure thing, my boy. It's a kind idea to bring a book to that poor cripple. I'll go a little later tomorrow so you can hitch a ride with me after your class lets out. See if Sister will give you early leave to do a work of charity."

Inga and Emmet cut one of the sketches out of her notebook and folded it inside an old copy of *Fireside Companion*. On the back of the sketch Inga wrote:

Dear Cyril,

Would you like to have your wood leg fixed like this? I can do it. Come to Ronan's barn alone after school some day. Tell us when you can come.

Your friend,
Inga Hagstrom

Emmet put it in his pocket. "I'll take it to him tomorrow."

Chapter 21
The Big Slice

That late February Saturday dawned clear, bright and cold. The snow left after the January thaw lay in thin patches on the ground. In the loft, Inga shivered as she pulled on her flannel petticoat and the brown wool dress. Wearing the new stockings Mrs. Duffy had made, she stuffed her feet into her boots and buttoned them up tight. She scrambled down the ladder quickly. "*God morgen,*" she told Mrs. Duffy, who smiled as she flipped pancakes.

"*God morgen* to you too. Now we've exhausted my Swedish and we must carry on in English." She loaded three pancakes and some bits of ham onto a plate for Inga. "Are you going to carry out your plan today? Are you sure you'll be safe? Mr. Brady is strong and has a wicked temper. Be careful."

Inga wasn't worried. "Cyril sent us the letter.

He says father has gone to Charles City to buy two sows. He will grow pigs in spring. He comes back late in afternoon."

"...and you are sure Jim Ronan will let you use his barn?"

"Yes. He promised Emmet. He said come early in morning. Emmet told his father he has to help Cyril. We will ride Emmet's father's horse." Inga gobbled the rest of her breakfast and went for her coat and shawl.

"Be careful," Mrs. Duffy said, giving her a little hug.

"We will be safe," Inga said. She patted her pocket where the hinges, springs, and screws jangled softly.

Emmet was climbing up on his father's horse as Inga arrived. He reached down to pull Inga up behind him and adjusted the saw where he had strapped it to the side of the saddle. Off they trotted towards the edge of town and the farms beyond.

Miss Mary open her kitchen door a crack and waved as she saw them nearing Ronan's barn. "It's good she knows we're here," Emmet said. "She'll keep an eye out for trouble." The barn door had been left ajar for them to slip in. Inside, the animals' warm breathing took the chill off the air. Ronan's big horse Belle neighed in welcome. They pulled a hay bale into the middle of the floor for a workbench, and lit a lantern.

Soon they saw Cyril working his way across from his own farm, avoiding the puddles covered with a thin skim of ice. He was limping on his crutches but had a big smile on his face. Once inside they set to work quickly. Cyril took off his wooden leg and laid it across the hay bale. Inga carefully marked the place for the saw cut. Emmet and Cyril held the leg firmly while Inga started the cut. Then they changed places and Cyril took a turn with the saw. He was older and stronger so his cut went faster. Soon they were more than half way through. "Bet I'm the only one in Floyd County to cut off my own foot," he laughed. "Doesn't hurt as bad this time either."

Suddenly there was a commotion in the yard and they heard Brady's voice. "Jim," he called. "Jim, c'mon out here and see the sows I just got. They're so plump already they'll give us good litters of piglets come spring." He waited to hear from Jim and the kids held their breath. "Jim! Are ye home?" he shouted. "I'll go 'round back by the chicken house if you're there." They heard him clump around to the back of the barn out of sight.

The kids looked at each other round-eyed. "He's early!" Run? Where? Hide? Under what? "Go to Miss Mary," Emmet whispered. "She'll save us." Inga stuffed the tools in a basket and Emmet grabbed it. Cyril Picked up his crutches. Inga picked up the leg.

"Go quick," Cyril hissed. "He's fat and slow. Even on crutches I can outrun him."

They slipped out of the barn and sprinted towards Miss Mary's back door. Inga got there first and pounded. Emmet pulled up next to her and helped her pound. "I'm coming! I'm coming!" they heard from inside. "Good gracious, what is going on out here?" Miss Mary opened the door and the kids tumbled in. They turned to check on Cyril—just in time to see Mr. Brady closing in on Cyril. His head was moving faster than his feet. He lunged to grab for Cyril's crutch, but slipped in the mud and fell flat on his face. Cyril hobbled up the steps to the Kelly house. He made it inside the door breathing hard, and flopped down on Miss Mary's floor. Miss Mary shut the door and locked it. "What in the name of sense are you doing to Cyril's leg?" she asked when she looked at what Inga had in her hand. "Jim said you aimed to fix it but it's almost cut through!"

"It needs an ankle," Inga explained, "so the foot can go up and down. So he won't be useless any more."

Brady was roaring outside the door, pounding and saying bad words. "We have a problem here," Miss Mary said. "Cyril, your father is upset and I don't blame him. That leg cost him a lot of hard-earned cash. Now it's ruined."

"No not ruined, Miss. See we cut this and put a hinge just here," Inga said, pointing. "Mr. Jim Ronan followed my model and it worked. We did all the testing. It will be patented and help the soldiers."

"I hear Jim now," Miss Mary said suddenly alert. "Let's hope he can calm our neighbor."

There was some quiet talking, then some yelling. Then more quiet talking. The yelling got softer. Then a key turned in the lock from the outside.

"Anybody home?" Jim's voice asked cheerfully, poking his tousled red head in the door.

"We have a visitor here," Miss Mary told him. He looked in to see Cyril sitting on the floor and Inga holding a partly cut wooden leg. Emmet was holding a saw and his wife Miss Mary was turning a hinge around in her hands to see which way the leg should bend.

"It looks like you're all busy here. Would you

like a helping hand?"

Miss Mary sat down on a bench and eyed her husband. "Are you part of this?" she wanted to know.

"I believe I am, though I didn't predict quite this way of doing it. I helped in testing. I can finish the cut if Mr. Brady agrees." He turned to Mr. Brady on the porch, who was angrily brushing the mud off of his coveralls. "What say you, Cletus? Will your son have an improved, never-before-seen invented bendable leg? Or a cut-up one that has to be replaced?"

"I guess new-and-improved is better than none," Brady growled. The men cleaned their muddy boots on the scraper outside, and stepped gingerly into the house.

Brady nodded a greeting to Miss Mary. "If it doesn't work I'll have the sheriff on ye, Emmet."

"But I never..." Emmet said. "I didn't..."

"The invention is mine," said Inga. "I have all of the tests. It will be for soldiers too. Cyril is the first. He will not be useless."

The adults looked at each other and shook their heads. The nerve of the girl!

Jim put the leg on Miss Mary's work table and Emmet and Cyril held it down. Inga showed where the cut should finish. Jim sliced through the wood, making Mr. Brady wince at the rasp of the saw. Jim made small holes in each section and screwed in the hinge just like they had done on the test leg. He screwed the springs in place.

Cyril rolled up his pants leg and fitted the leg onto his stump, tightening the straps to hold it in place. Jim gave him a hand up. Cyril took a couple of steps, working his "ankle".

"If that don't beat all," Mr. Brady said, shaking his head in disbelief. "The boy is walking. No crutches." He pulled out his big kerchief, wiped his eyes, and blew his nose. "I never thought I'd see the day that boy would walk again. You kids did good."

Chapter 22
The Letter

On Monday morning the class was abuzz with excitement. "He done it again," one said. "That Emmet is crazy but he comes up with some great ideas."

"Imagine putting a hinge in a wood leg just like a door. Who would have thought of that?"

"Did you see Cyril Brady walking up proud at church?" one of the girls chattered to her friends. "He walked right up the communion rail like a man. And he smiled at me going back..."

On and on the chatter went, until Inga came into the room. "What do I hear of Cyril's leg?" she wanted to know. "And what of Emmet?"

"Oh, he invented a better leg and now Cyril is walking."

"It is my invention," Inga corrected. "It is my idea. Emmet only helped. It is an invention and people will have to pay me to copy it."

"Any fool can put in a hinge."

"They can put it in but they must pay for my idea," Inga argued.

Emmet walked in just then. "If she gets a patent, the idea's all locked up. I looked it up," he said. "Better fix all your legs mighty fast before Inga gets the paper from the government. I found it out. It's the law." He smiled at Inga. "You're my smartest partner," he said.

After school Inga stopped at the post office to pick up the Duffy's mail. "How long does a letter take from New York, Mr. Roche," she asked. "I hope I get a letter soon from my papa."

Hiram Roche proudly pulled a letter out of the sorting rack. "Here it is, Inga! Just in with a postmark New York. It's finally come." Inga tore the envelope open and her face fell.

"This is from Mrs. Hobson. She wants a report in fall. There is nothing in the letter about my father. She doesn't know where he is. My father is lost for me." A small tear leaked down her cheek. She brushed it away as if it was a bug and shook her head. "Now I have nobody."

Emmet came out of his workshop in the back room. "I thought you were here, Inga. I heard your books crash down. Why are you crying?"

"Not. Not crying." She sniffed hard and brushed at her eyes. "Letter is from Mrs. Hobson, not father. Letter is not good."

"Are you arrested? Are you in trouble?"

"No," she almost shouted. "I only want a family, not a letter from Mrs. Hobson." She crumpled the letter in her hand.

"You have us, Inga, don't forget. You have Mrs.

Duffy and Flicka, and Tom and me and Sister and Big Jim and Miss Mary and Jim's horse Belle."

Inga looked seriously at Emmet. "All those are my family?" She started to laugh. "It's a funny family, but better than nothing. You and the horse will have to be my cousins, then. Let's watch for more letters. Something will come for sure." The big clock struck four loud bongs. "I have to go back to help Mrs. Duffy." She stuffed her letter into the pocket with her tools, picked up her strap of books and marched out.

Emmet and his pa Hiram looked at each other, shaking their heads. "She's a strong one, that she is," Hiram said.

"And smart," Emmet added. "She's almost as smart as me."

Chapter 23
The Goat Cart

"That Eamon McNally gets me sick," Tom sputtered one day in late spring, leaving school. "He gets to have a new suit every Easter and play pool in his dad's saloon with the men. Now he's bragging that his father is getting him a goat cart."

"Does he have a goat?" Inga wondered.

"He's getting one of those too. And a new cart that he can drive like a real wagon! It's painted bright shiny blue and the spokes on the wheels are red. He really thinks he's something. What a fourflusher!"

"Goats are ornery," said Emmet. "Not worth the bother. Driving a goat would be like driving a cat. It wouldn't behave at all."

"It might be nice to make a cart that would go without an animal," Inga suggested. "Maybe put parts together so the cart could drive by itself."

"Like Benz?" Emmet suggested. "In Germany

a man named Benz invented a car that goes without a horse. It has a motor. It would be grand if you could figure that out!"

"How does it work?" Inga wanted to know. "It must burn something to make power. Like the trains burn coal for steam."

"I read it has an engine and it burns gasoline."

"What's gasoline?" said Tom.

"Sort of like oil but juicy. It heats up in the engine and makes things turn. They make it out of oil and it can explode. I read about it," Emmet said.

"My father helps make locomotives," said Inga. "No horses there. Steam is best. It can move anything."

"Can you make one, Inga? Something better than a goat?" added Tom. "It doesn't have to be as strong as a train."

"I'll think about it," she said. "But you'll have to find a play wagon." They walked on, enjoying the first whiff of spring flowers blowing in from the prairie.

When Inga got home, she began drawing right away. She drew pulleys—too slow. She drew gears—too expensive. She drew a bunch of levers—big wobbly mess that folded like grasshopper legs. She tore the pages out of her notebook and stuffed them into the stove. One day, while sweeping the kitchen, she noticed Mr. Duffy's belt hanging on a peg. A belt, that's it... or maybe two. A wheel could turn the belt and the belt could turn the wheels.

She remembered a time when Papa described the big belts that power saws to cut trees into lumber. He had said they were so big they could reach across a room. Unfortunately, Mr. Duffy wasn't that big even with all the pies he ate. But the idea stayed in her mind. *I wonder where we can get a big, strong wheel* she thought, *and handles to turn it. Oh yes, we also need something like a thread spool for the back axle.*

A few days later, Inga saw Eamon outside trying to train his goat. The goat listened patiently, chewing a mouthful of grass. It blinked its beautiful big eyes at the children looking on. Then it turned and tried to take a bite out of Eamon's straw hat.

When Inga got to school, she asked Tom, "Did you get the play wagon? There was one in the general store when I was working there before Christmas. I don't remember if anybody bought it."

"I asked Mother. We have to buy it though. Mother doesn't give things away if they're good."

"I have some ideas and drawings. We'll need some more parts too."

They joined the crowd of children going into the school as the big bell rang out.

While Inga and Emmet waited for Sister to come in and start the day, she continued, "I have made many drawings but I cannot draw my whole idea. It is time to put the parts together. We'll see if it works. Do you want to be on my team?"

"I can do all the research. I'll find out more about that man Benz..."

"No. I need a team that will measure and cut. We need to put parts together. I need thinkers, not readers."

"I'm a thinker too," said Emmet, slightly offended. "Reading doesn't stop thinking."

"Yes. That is true, but this idea must come from our own heads. Nobody has done it yet."

Sister arrived and everyone stopped talking.

She tapped her desk for attention and the lesson began.

At recess, "I know where we might get a free wagon," Tom said. "Mother has an old one in our cellar that somebody traded her for a bag of seed corn. It has a few broken parts. If she can't sell it I'm sure she'll be glad to be shed of it."

"Does it have four wheels?"

"Yup."

"Handle to steer?"

"Yup."

"Does it have sides?"

"Yup... but one is cracked."

"Is that all that's broken? There's not much to a wagon," said Emmet. "Sounds good so far."

"The paint is worn off in places too."

Inga thought for a minute. "Easy to fix. We must get more parts, too."

"Like what?"

"A wheel for power, pieces of wood, and some, what do you say,—*skruv*?" She moved her hand as if using a screwdriver.

"Screws! Almost the same," Tom said. "We can get that stuff. Maybe Mr. Duffy has them from fixing wagons. There are always pieces left over. You know that big box where you got the hook to pull up the flour?"

"Yes. Many parts there."

"There will probably be screws in there," said Tom.

"All this about parts is nice, but how do you plan to get it to go?" said Emmet. "That's where you need the research."

Inga pulled out her notebook just as the bell

rang. She snapped it shut. They hurried back inside.

"You three look happy," Sister Norah James said. "Are you getting spring fever?"

"We're thinking," Emmet said. "Thinking of great ideas. It's fun."

Sister looked doubtful. "It's nothing dangerous, I hope."

"Not us, S'ter," they replied.

Chapter 24
The Wagon

On Saturday afternoon while Inga was finishing her chores, Tom delivered the old wagon to Duffy's. He rolled it under the lean-to next to the washtubs, and knocked on the back door. "Here's the wagon. What do you have in mind, Inga?" said Tom. "I want to drive it, at least for the race against Eamon and his goat."

Inga hurried outside, wiping her hands on her apron. She pulled her notebook out of her pocket.

"I think we can put a strong wheel here," Inga said, pointing to the middle of the wagon. "You will sit here at the back. You will steer with this thing." She held the long handle and moved it from side to side. The front wheels moved right and left.

"What's another wheel going to do on the top? Is it in case I roll over?" Tom teased.

"It moves the belt."

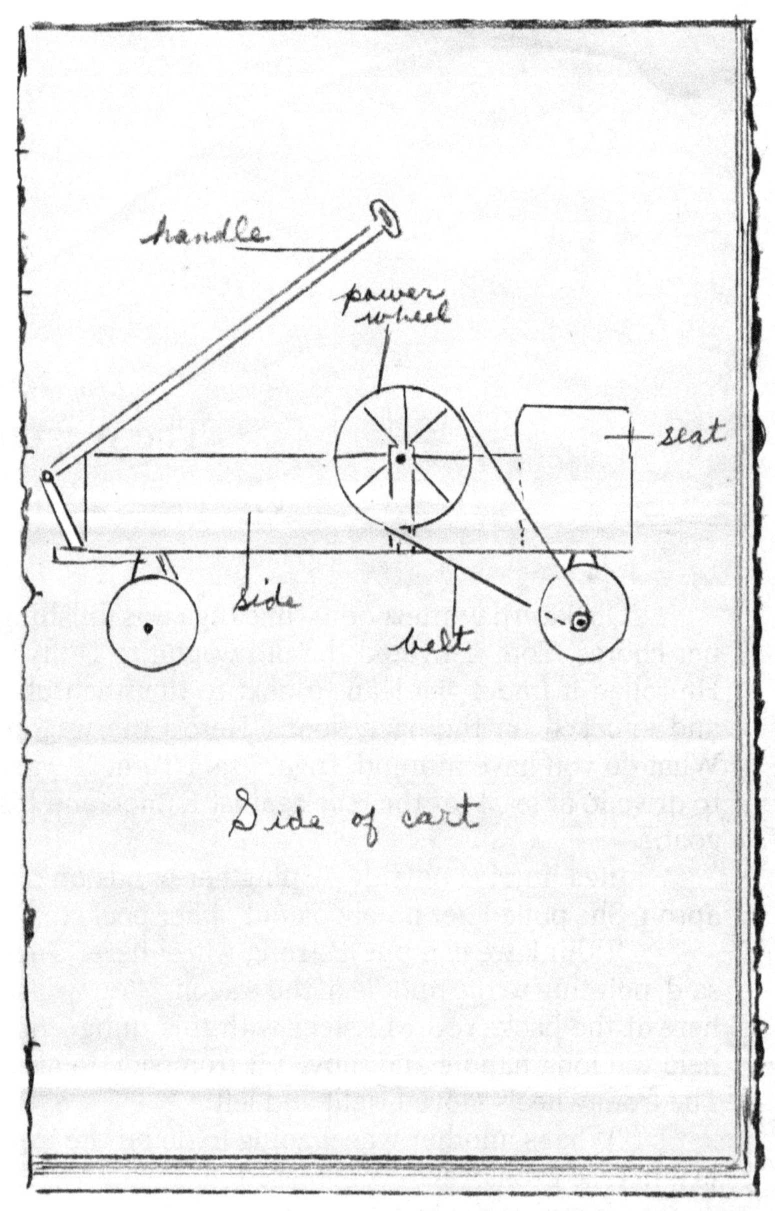
handle
power
wheel
seat
side
belt
Side of cart

"What?"

"The belt goes around the wheel, down to other wheel, and makes the cart go."

"How do I move the big wheel on top?"

"With your feet, I think."

"I'd think you'd want the feet down near the ground," Emmet said. "Put the machinery down there."

"Show us," said Inga.

Emmet pointed to where Tom's feet should go, and then realized there would have to be holes in the wagon floor for them. And there wasn't much room down there either. "Hmm. I see why Benz used an engine."

"If you crash with your feet down you might get hurt like Cyril," Inga said. "Better keep your feet up."

Tom had been thinking, walking around the wagon, stroking his chin. "What holds the big wheel up? I was thinking of an axle across from side to side, but my feet would get caught the first time around. It just won't work. Wheels can't stand up in the air."

"I was thinking about that. We can make some kind of thing to hold up the wheel. Make it strong, but without a long axle across it. The feet will have to push it. The belt will go between the feet, from the top wheel to the wheel at the bottom."

"I see," Emmet said. "Feet up, belt down. We have to make a hole for the belt then."

"Yes, a hole for the belt, but we have to keep the wagon floor strong."

"A small hole," said Tom.

"No, a long skinny hole, just for the belt," said Inga.

"Do you plan to get a belt from Duffy?" Emmet giggled. "He's about as big around as we'll need."

"Maybe. Or maybe cut an old horse —". She shook her hands as if she were driving a horse wagon. "What do you say, for horse."

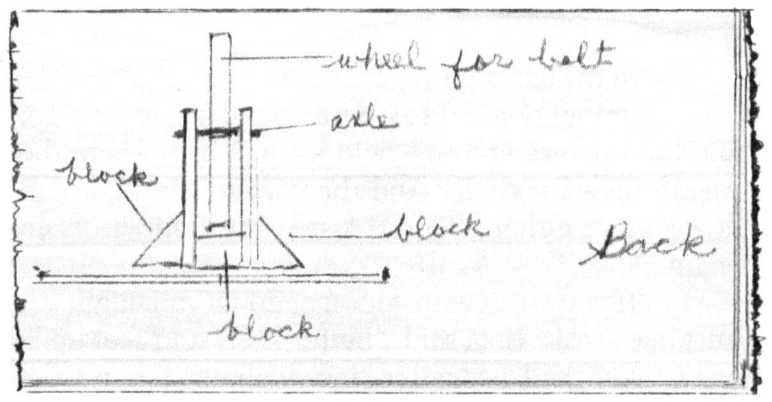

"Handles? No, reins," Tom realized. "Yes, that would give a nice long leather piece. We can glue it to make a belt of the right size." With farmers going in and out of his mother's store, Tom was sure he could find someone with a spare to give or trade.

They gathered around Inga's sketches. She had drawn two sturdy pieces of wood with a smaller block of wood between them. She added heavy triangles of wood to the upright pieces so they were strong enough to support the wheel. At the top, a groove held the wheel by its axle so it could spin. The structure was screwed into the wagon floor to hold the wheel up between the driver's feet.

Tom was the first to notice a problem. "What do I put my feet on? Something like stirrups so I can turn the wheel."

"I don't know yet. We don't have a wheel for this." They were stumped and just sat there for a bit, thinking. "Let's pretend," Inga said finally. "Tom, sit in the wagon, at the back." He did. "Now stretch your legs out. See how far they go." He did that too. Inga made a chalk mark on the side of the wagon. "Now fold your legs up. Not too tight." He did it. She made another chalk mark.

"Paint would work better," Emmet said, joking.

"I need to see how the big wheel works for Tom's legs. A big wheel will give it more power, but if it's too big, he will not be able to turn it."

"Yeah, we have to make it fit for me," said Tom.

"Yes. Will you measure please?" She had made another mark halfway between the first marks. Emmet got the yardstick he had borrowed from his father and measured. It was 14 inches between the first two marks.

"What an eye you have, Inga! Fourteen inches and the middle mark is smack at the seven."

Inga explained, "The middle mark is for the place where the wheel holder goes. Tom can stretch fourteen inches, but that's too hard to pedal. A 10 inch wheel will give him room to move his legs easily."

"Where are we going to get a ten-inch wheel?" said Tom.

"How 'bout your mother? Doesn't she have spare cart wheels?" Emmet said.

"If it's new we have to pay for it."

"Mr. Duffy?"

"I have to ask Duffy for a *skruv* and maybe an axle," Inga said. "And I didn't see any wheels in the shop."

"Sisters won't have wheels."

"Don't think so." Both boys shook their heads sadly. "Sister Ludwig has lots of junk, though."

"We must find a *tygel* too—you say the word is 'reins'? Maybe a farmer…"

Finally Emmet said, "Let's all ask everyone we know for an old wheel about this big," he showed with his hands, "and one old spare rein. Does anybody around here have a new set of horse harnesses? They may give us a piece of their old ones." Nobody had a suggestion.

"After we get the parts, we can start making it," Inga said. "I shall make better drawings while we find the parts." The church bell rang. It was six o'clock. Soon the clanging of dinner bells signaled time for supper, ending the project for the day.

Chapter 25
Goatless Cart

It was two full weeks before they could meet again. Tom had begged a small carriage wheel from Sister Ludwig's spare parts collection, and one slim piece of old leather 12 feet long from a liniment salesman passing through town. Inga had convinced Mr. Duffy to give her several carriage bolts and four sturdy wood screws. Emmet had wheedled two old stirrups from Cyril's father, Mr. Brady, who said, "at least you can't do any harm with these."

They worked until sundown, and started again on Sunday right after Mass. Mrs. Duffy was concerned with their working on a Sunday but Emmet assured her that it was fun and educational, not work. Inga and Emmet cut the longer pieces of 2x4 to size and screwed them together with the shorter piece. They carved grooves on the top of the long pieces to carry the wheel's axle. Then they glued and screwed the triangles to sides at the bottom. It was beginning to look like a frame, almost like Inga's picture (see p.156). They drew a chalk line straight across the wagon at the center mark and set the frame on it. They balanced the axle and wheel in the grooves and stood it up in the wagon. It was beginning to look right. "Here, Tom, try and sit inside," Inga said. The cart crashed down as Tom started to climb in, knocking off the rear wheel and axle. He propped the cart back up with a piece of scrap wood and put the frame and wheel back on.

"Don't know as I can sit in it, Inga," Tom said. "Don't want to crash it yet."

"Just stand next to it, then. We have to see if you'll fit."

Emmet and Inga stood back and squinted at Tom every which way. "I think he will fit," Inga said. "But he sits too low. We need to find a box for a seat."

Tom went back to his axle work, threading a wooden spool on the center of the rear axle and gluing sandpaper around it. That should grab that leather belt good, he thought. I'll go like lightning.

Emmet and Inga sawed a long narrow hole in the floor of the wagon for the belt to go through. Then they set out to attach the stirrups to the wheel for Tom's feet. They tried glue—didn't work. They tried screws— too hard. They thought about using wood but couldn't decide on the shape. Finally Inga remembered her carpetbag handle and the way she had repaired it. "Go get a spool of thin wire," she ordered Emmet. "If Mrs. Duffy doesn't have it, ask Tom's mother in the store. Say I will pay for it."

Emmet tore off around the corner of the house and down the street. Inga waited, turning the stirrup this way and that to imagine how the wires should go.

Soon Emmet was back with a small roll of wire. Inga used it to weave a connection between the stirrup and the wheel. She neatly anchored each stirrup to a spoke so it wouldn't slide, but could turn with Tom's foot.

Tom had the rear wheel back on and the cart stood straight. They put the frame up on the chalk mark and bolted it through holes in the wagon floor. "What about the belt," crowed Emmet. "We forgot the belt."

They used a long piece of string to measure around the power wheel, through the slot, and around the spool Tom had put on the axle. They laid this out on the leather strip and cut the leather two inches longer so it could be glued into a loop.

While the glue dried they went into the house to ask for a snack from Mrs. Duffy. The pie and milk were so good. They stretched out on the floor to rest and then woke with a start. It was almost dark and they hadn't finished. Mrs. Kelly was ringing her dinner bell loud and clear, and Mrs. Duffy was stepping over them as she put supper together. "I think that you've finished for the day, children," she said. "You can finish after school some time, or next week."

"But what if it rains?" they complained.

"Push it deep under the roof edge. You can keep the leather in here to dry. Wet won't harm the wood any."

"But all the parts..."

"Any loose pieces can be here with the belt. No harm will come to it."

"Let me have a look," Mr. Duffy said, getting up from the rocker. "If the weather breaks up I'll put it in my shop and anyone looking to harm it will have me to answer to." They all trooped out the back door

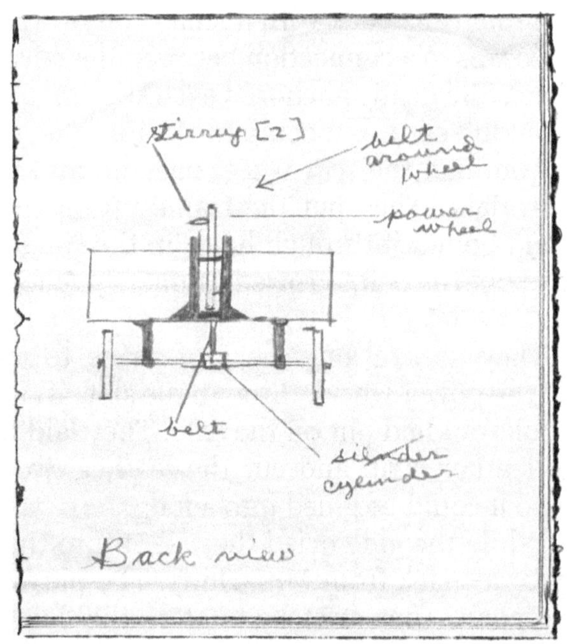

and looked at the cart. Even with some parts off, the plan made sense to Mr. Duffy and he said so. "It's a clever idea," he said. "Never expected it to be so well-planned. You do have a problem, though, and I'm not the man to tell ye. You'll find it yourselves. Now I bid ye a good night." He waved the boys away and turned back into the house.

Inga had trouble sleeping that night. What did Mr. Duffy see that she couldn't see? What had they left out? They measured the belt, and glued it tightly. They wired the stirrups on the wheel. The wheel fit in the frame. They bolted the frame down tight. They put the back wheel in place. She turned over and over until Flicka climbed up on her chest and snuggled close. "Time to sleep, little one," she mumbled in Swedish as she fell asleep.

Chapter 26
Building the Cart

That week it was hard to pay attention in school. The stories in the reader seemed dull. The arithmetic problems were confusing. Even jump rope rhymes had lost their interest. *What did I miss?* crowded out Inga's other thoughts.

Finally on Saturday the boys showed up right at noon. "Ya done working yet, Inga?" they hollered outside the back door.

Mrs. Duffy poked her head out her door. "Inga is just finishing her tasks, boys. You may go into Mr. Duffy's shop and retrieve your materials quietly. Don't disturb him. He's concentrating on an ornamental piece for Mrs. McNally's front door. You know how fussy she is."

They tiptoed into the shop. Mr. Duffy was beating a red-hot piece of iron into a curlicue. Sparks flew in all directions, bouncing safely off Mr. Duffy's leather apron, and narrowly missing the wheel, frame, bolts and leather piled up in the corner. Duffy nodded hello to them, never taking his eyes off the glowing iron.

When they met Inga outside they laid out all the pieces in order. "Any ideas?" said Tom.

"Lots, but they're in books," Emmet admitted.

"If we put it together in order, the belt will not fit around the axle," Inga said. "Look." She put the belt around the power wheel, slipped it down through the slot and showed them the problem. It couldn't go around the axle. They got down on their knees to peer under the wagon. Tom reached in and slid the belt along the axle. "If I could just get around..."

"That's it," Inga cried. "That's it, Tom. Take wheel off again. Move the axle to the middle. Only to put the belt around it, then move it back. Put the wheel on again."

He did that but it wasn't easy putting the wheel back in place.

"It's too tight, Inga. Lots too tight."

Inga squinted under the wagon to see how the parts came together. "Can we make the spool smaller? We need a file."

Emmet borrowed a file from Duffy. They slipped the belt off only a bit to the side, and filed the spool thinner all around. All of the filing ripped some of the sandpaper off, but the belt gripped like crazy now, once it fit.

"You try," Inga said to Tom. "You are the rider."

Standing at the side, Tom held the stirrup with his hand to turn the power wheel a little. The back wheel slipped a bit, then started to roll forward. They took turns pushing the stirrups with their hands to make it go. It seemed to work. "Get in it now Tom and use your feet," suggested Emmet.

"We need the seat first," Inga reminded them. "It's too hard to work your legs if you sit on the floor."

They found an empty crate the right size and screwed it into the wagon. Tom climbed in and sat on the crate. He put his feet in the stirrups. He held onto the steering handle. Carefully he pushed the stirrups with his feet and the cart moved. It really moved. "Hooray," he shouted.

"Eureka," Emmet crowed.

"I am glad," said Inga. "The idea was good."

Chapter 27
The Dare

The next afternoon Eamon McNally was already waiting outside school when classes let out for the day. He leaned proudly against the side of his new, freshly-painted, blue cart. "I heard you have a cart without a horse," he snickered. "Want to race? William here has been lookin' for a challenge." He rubbed his hand along his goat's smooth horns, polishing the shining brass balls on the ends. Then he patted William's nose. He adjusted the bright red harness strap making sure everyone got a good look.

Before Inga could say a word, Tom stepped forward, his chin out in defiance. "When and where?" he challenged. "Inga's cart don't need feeding, so we can run it whenever we want to. Don't have to clean up after it, neither."

"No school tomorrow. Just bring your little girly invention over near the church in the morning.

We can run down the street to the Widow Doyle's, where the street is smooth. It'll let William get up more speed. He likes a nice flat track—easier on his hooves. What say 10 o'clock?"

"I didn't make a race cart," Inga protested. "We didn't finish testing it."

"Well, if you're scared to race..."

"We're not scared," Tom said. "Our team has the best cart and you'll see it tomorrow."

By this time a small crowd of kids had gathered to hear the discussion and started to sort themselves into two groups, each rooting for its chosen team. They kept on arguing among themselves as they scattered towards home.

After supper the boys came to Duffy's shop to help Inga with final adjustments. Tom helped Inga remove the rear wheel and axle, and glue new sandpaper around the axle to get a grip on the belt. They glued sandpaper all around the rim of the drive wheel too. Emmet tightened the belt and put grease around the moving parts. "Don't get it on the belt!" Inga squealed as Emmet reached in with the pot of grease. "Not a drop. We'll lose all our power."

"Does it need a name, do you think?" Tom asked.

"Carts have names?" said Inga. "That's silly."

"It's like a team," Emmet answered. "Kids have to have something to shout about."

"We could name it for our town," Tom said. "We live in Floyd. How about Floyd Flyer?"

When Mrs. Duffy came with a kerosene lantern to tell them to stop for the day, they were ready to give the Flyer a last pat and go to their homes and loft to rest up.

Chapter 28
The Race

Morning dawned bright and dry with a light wind. The sun had dried the puddles the day before and the street was fairly flat. Kids frolicked along Main Street heading toward the Widow Doyle's place. She was famous for her chickens, Rhode Island Reds, that laid lots of eggs and made good eating. Every year, almost, she brought home the prize ribbon for Best of Breed chickens.

The kids lined up there, on both sides of the street, waiting for the challenge. By 9:45 Eamon showed up, sitting proudly in his cart behind William, who was munching on hay. Inga's team came up the street and parked on the other side.

"We have to have a start and a finish or it won't be fair," Emmet declared.

"We can mark it but you'll never get there," Eamon said. "Let's draw a finish line here in the dirt, from the chicken coop to the other side. We should start uphill by the church." He scratched a finish line all across the street with the handle of his buggy whip.

Everybody trooped back up the slight hill to St. Patrick's. The teams got their carts. Eamon drove William up to the church with great flair, waving his little whip and telling William to "git along." Inga's team pushed their cart back up the hill. Tom walked beside to save his strength. As the official driver, he didn't want to start out tired. It wouldn't do to lose out to a goat.

Sister Ludwig came out of the church and caught sight of the contest. "Good morning, children," she called. "Why are you all here? There's no school today, you know."

"It's for a great race," Emmet announced. "Goat against machine. Inga figured out how to make a cart that goes with nobody pulling it."

"Do you need someone to say 'Go'?" Sister asked.

"Wave your handkerchief up and down like a flag. We're starting here," Emmet said. "Hold it up high and when you say 'Go' pull it down fast." He pulled gently at Sister's sleeve to guide her into position.

Everybody lined up. Eamon got into his cart and maneuvered William over to the starting line. Inga and Emmet pushed the Floyd Flyer into place and Tom climbed in.

"Ready?" Sister said. "Set." She paused and looked over the little crowd. "Go." She brought her handkerchief down with a swish.

Tom held the steering bar tight and worked his feet in the stirrups to turn the power wheel. He started rolling. Eamon and William began to trot, easily passing Tom, who struggled to keep the power wheel moving. The supporters of the Floyd Flyer team groaned and the William fans cheered. One of them waved a leafy little apple branch like a flag. William spotted it and stopped to have a bite. Tom passed by, weaving from side to side. "Eat later, you dumb goat," shouted Eamon. William calmly munched on the rest of his snack and turned to gaze at Eamon. Eamon had never shouted at him before and he was puzzled. "C'mon you stupid billy goat, we're in a race." Eamon waved his little whip in the air and William watched it swing, chewing contentedly on his branch. Tom kept turning the power wheel, his boots almost sliding out of the stirrups. Finally William decided to run and almost caught up with Tom.

At the crest of the hill gravity took over. The Floyd Flyer rattled on ahead, passing the finish line about two feet ahead of William. Tom turned to say "Ha Ha" to Eamon and forgot to steer. The steering handle wrenched out of Tom's grip and swung wildly right and left as he hit bumps in the road. His left foot slipped out of the stirrup. The crate seat bounced around. Tom bounced too.

On the Flyer went, smack dab into Widow Doyle's chicken yard. Rhode Island Reds flew out over the street, squawking and cackling in panic. The Flyer came to rest against the wall of the chicken coop, covered with feathers and chicken droppings. William kept going, toward the edge of town, still munching on his apple branch.

Widow Doyle came out of the coop where she was collecting eggs and looked for somebody to blame.

The bystanders had vanished, and so had William, who paid no attention to the shouting Eamon. "You will find my chickens or buy them," she declared. "And fix my fence."

Tom, Inga and Emmet bowed their heads, spitting out feathers. "We'll catch them." They spread out in the area, sneaking up on the chickens one by one and carrying them back to Widow Doyle.

"Just a few little problems," said Emmet, pulling dark red feathers out of his hair and straightening his spectacles after he deposited his last angry hen in the coop. "It worked in principle."

"We forgot to put on a brake," Tom mumbled, rubbing his sore shin. "There's your principle. Whose bright idea was it to build a fast cart with no brake?"

"My mistake," Inga said. "I was always working on power. We weren't ready to race." She walked on holding on to the cart with one hand as the boys pushed. "I wonder if your man Benz had to deal with chickens."

Chapter 29
Inga's Father

The June sun shone in a clear sky and the scent of newly cut hay drifted over the Floyd train station. Mr. McGillicuddy, the stationmaster, leaned his head on his hands clasped on his walking stick as he sat on the bench waiting for the noon train. Hope there will be some mail for those kids, he thought. They've been at me every day to see if there's a letter from Inga's pa in the mailbag. Eyes closed against the sunshine, he snored softly. A fly landed on his ear. He didn't notice.

Two long toots from the engine made him blink. Must've been nodding off, he thought. The train rattled and squeaked, giving a last "chuff" as it let off steam and came to a full stop. A tall stranger with a short, neat, white-blond beard stepped down. He carried a carpetbag and had a violin case tucked under his arm. "Howdy," said Mr. McGillicuddy. "Lookin' for somebody?"

"Yes. Welcome. Thank you." He tipped his well-used hat and pulled an envelope out of his coat pocket. "I look for Duffy. Where is Duffy?"

"Duffy is our blacksmith. His house is two streets over." McGillicuddy waved across and back toward the town. "Go two streets, then right and you'll see it once you pass the church. It's the white house with a workshop on the side. Just wait a minute." He sat back down and pulled the mail sack over. "I may have letters for that house." He dug down into the bag beside his chair, looking at each address. "Nope. Nothing today. The kids have been asking every day about a letter from Washington. Say, what did you say your name is? You look familiar somehow."

"I am Hagstrom, Alex. I came to find my daughter Ingeborg. She works for Duffy--"

They were interrupted by a loud commotion. All the way from Iowa street, they heard hooting and hollering. Children were cheering and laughing.

"Sounds like a parade," Hagstrom said. "It can't be somebody running for governor. It's only June." They walked together to the end of the station and squinted toward the noise to find its source.

"Oh, it's only the kids," McGillicuddy said. "It looks like they have that mechanical wagon working again. It was laid up for a week or so after the crash into the chicken coop."

"Crash?"

"They raced a goat cart on a dare. Ran into Widow Doyle's chicken coop. There was lots of excitement, but the goat lost. We were picking up chicken feathers for a week."

Soon the little parade came back into view.

Alex Hagstrom watched intently this time. Yes. That girl looked like Inga. Same hair. Tall and thin. She was trotting beside a shabby cart. The boy seated in it pedaled furiously on what looked like a carriage wheel. A cat was riding with him. Alex waved vigorously, hoping to catch the girl's eye.

When they were less than a block away, Inga noticed someone waving, and could hardly believe what she saw. Her father had come! There was no letter. The train had delivered a real person. She broke away from the group and ran as fast as she could toward the station. "Papa! Papa, you've come," she cried as she ran, followed by the crowd with the cart. They surrounded Alex on the platform, almost knocking him off his feet. Inga threw her arms around him. "I'm so happy, Papa. I worried you might forget me."

"Never," he replied.

"Careful, children," McGillicuddy shouted. "Get back from the tracks. The train is about to leave." With a couple of long bursts of steam and the metallic screech of wheels, the train slowly started to roll, gathering speed as it chugged northward out of town.

As the children backed away from the train, Alex Hagstrom had a chance to look at Inga carefully. She had grown tall and filled out a bit, but that was the least of it. She glowed with confidence and carried herself proudly. "I could never forget, *dotter*. I was working to save up and move west. Tell me about your life these last months. I didn't get any letters."

"But I wrote to you. I wrote twice. I wanted you to know I was safe."

"I never heard from you or Mrs. Hobson.

Maybe it was because I moved. I got a bigger room in a place that catered to more decent roomers. Mrs. Lundgren didn't keep an orderly house. But tell me about yourself, now. How have you been keeping?"

"I have been working: learning English, making pies, finding friends, making inventions."

McGillicuddy butted in. "She looks like an angel, until she pulls the pliers out of her apron. Then you want to watch out for your machinery."

Alex Hagstrom smiled broadly at this. "Do you mean that cart?" he asked. "You have quite a crowd of admirers. I'd like to see how the cart works."

The boys rolled the Flyer proudly closer to Alex, and Inga introduced them to her father. "This is Emmet and this one is Tom. They help me make my ideas."

"We get a penny for a ride. Or a lemon drop," Tom offered. "We have a business."

Alex leaned down to look closely at the cart. "I see, you used a belt drive. They slip, you know." He looked at their fallen faces. "But a good idea if you can control the slippage." He looked more carefully and saw the sandpaper around the rear wheel. "Good idea there." He felt the rear cylinder and tapped on the belts. He looked at the wiring that attached the stirrups. "Your handiwork, I see, *dotter*." Inga nodded.

"The whole idea was Inga's," Emmet explained. "We helped put it together."

"And tested," Tom said.

"I'd like to see it go," said Mr. Hagstrom. "See it in action."

Tom climbed in, settling himself on the seat crate. "I'll ride it here next to the station where it's flat. No need to try the hill." Tom set off down the street pedaling away. Alex Hagstrom turned to Inga, covering a smile. "Did you forget anything, Inga? There seems to be something missing."

"I'm thinking about a brake. We had troubles at the race. Did the stationmaster tell you?"

"He did. He thought it was very funny. But Tom is driving it around without any trouble."

"When the street is flat, he can stop pedaling, or pedal backward a little. Stopping is only a problem on hills." Inga smiled up at him. "Now you are here. You can help us with problems."

"I think you can think these things out for yourself, daughter. You've accomplished a lot on your own."

"Will you live here now, Papa? You can stay with me. I don't know if Mrs. Duffy will let you stay in the loft. It is full of flour."

"We made a promise, remember, Inga? I signed for you to stay and help Duffys until you are eighteen. I need a job myself. I have one waiting for me in Waterloo, designing farm machinery."

"Waterloo! Mr. Duffy goes there to get tools sometimes. It is not as far as New York, only two days. He goes with his wagon on the road. Maybe I can come and see you."

"That is a great idea, Inga. We can see each other and still have our jobs. Time goes so fast. You are already almost a young lady. Twelve years in June, isn't it?"

"Yes. Mother called me her *midsommar barn* because my birthday is on the longest day."

"Ah, I remember as if it was yesterday. And now I see you becoming a tall, smart woman."

"Oh, Papa. Not yet, but I will have my own life, won't I? After I finish my work for the Duffys?"

"Of course."

"I will decide on my own what to do."

"Yes."

"Then, I want to go to college to learn to be an inventor. I will save up for it. I already have three dollars and 12 cents."

"You seem to have made a good start, *dotter*. That is worth saving up for."

"Let me show you where I live now, Papa."

"I'd like that. Show me the way."

They all trooped off toward the Duffy house. Inga led, swinging her father's hand and carrying his violin. Emmet skipped alongside the cart, warning Tom of bumps ahead. And Flicka rode proudly in the Floyd Flyer, finally at peace with Tom. "You'll love the workshop here, Papa," Inga said. "There are so many tools. A person can make anything."

Sample Patents

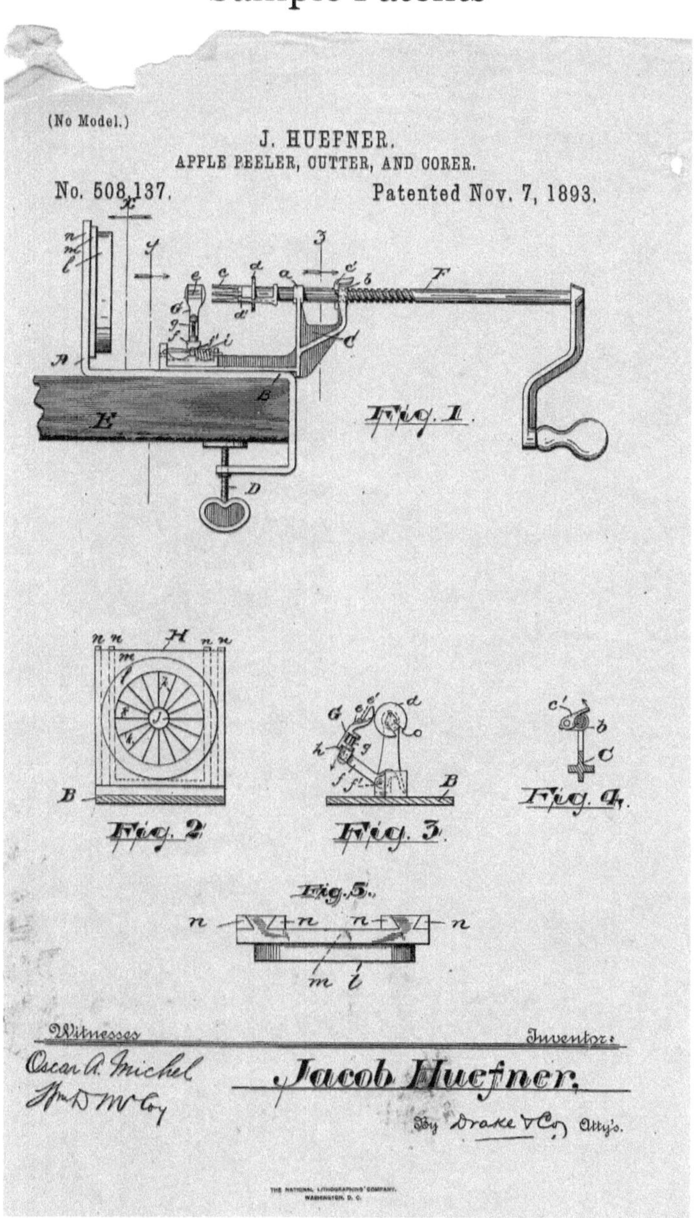

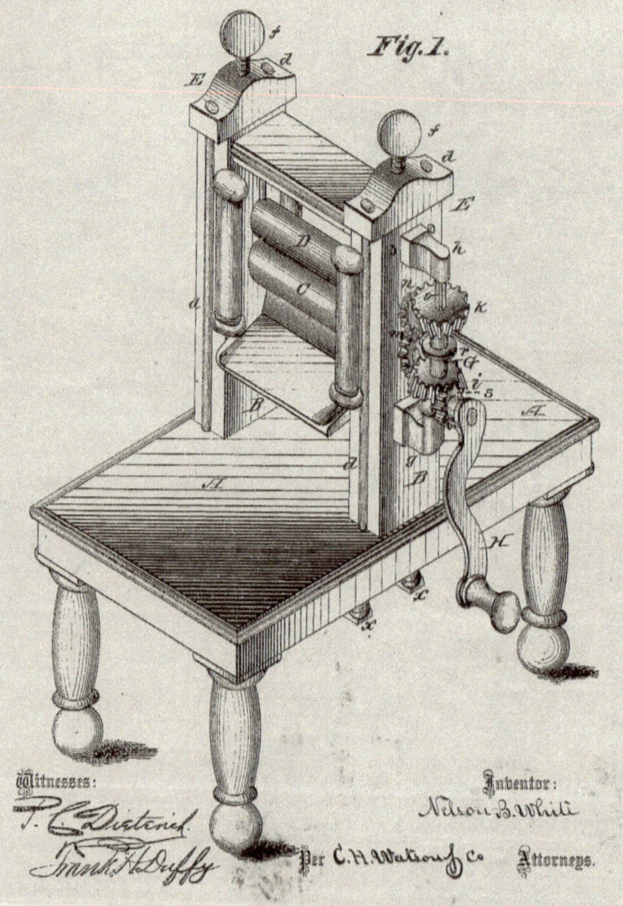
N. B. WHITE,
Assignor to Empire Wringer Co.
CLOTHES-WRINGER.
2 Sheets—Sheet 1.
No. 7,761.
Reissued June 19, 1877
Fig.1.
Witnesses:
Inventor:
Nelson B. White
Per C. H. Watson & Co   Attorneys.

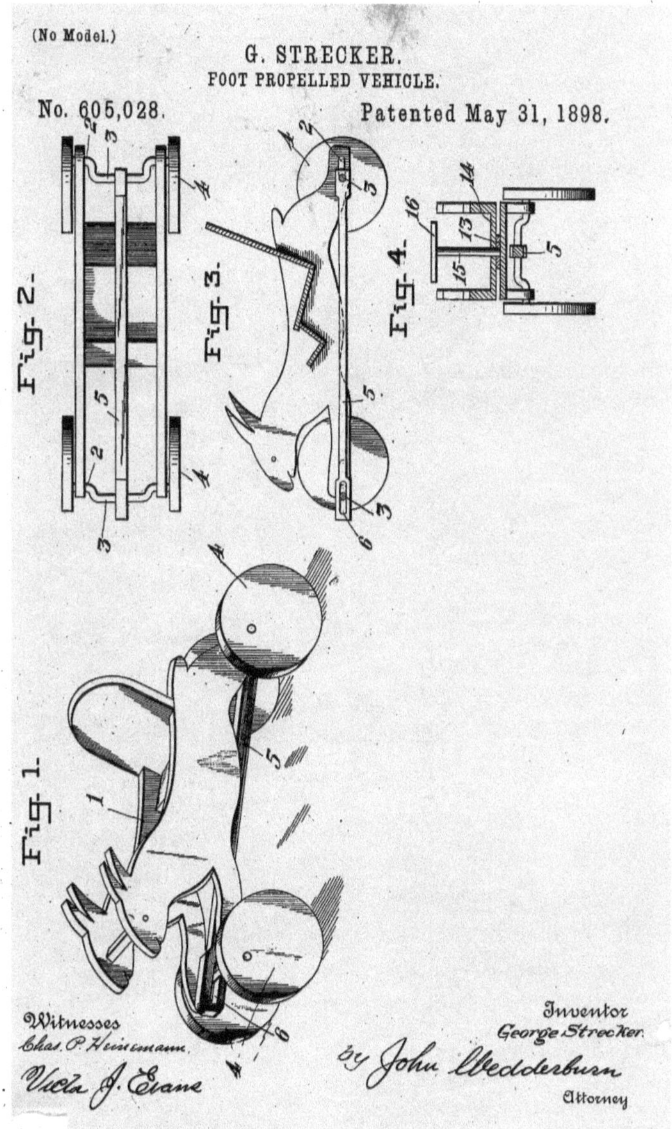
(No Model.)
G. STRECKER.
FOOT PROPELLED VEHICLE.
No. 605,028.
Patented May 31, 1898.
Fig. 2.
Fig. 3.
Fig. 4.
Fig. 1.
Witnesses
Chas. P. Heinemann.
Viola J. Evans
Inventor
George Strecker.
by John Wedderburn
Attorney

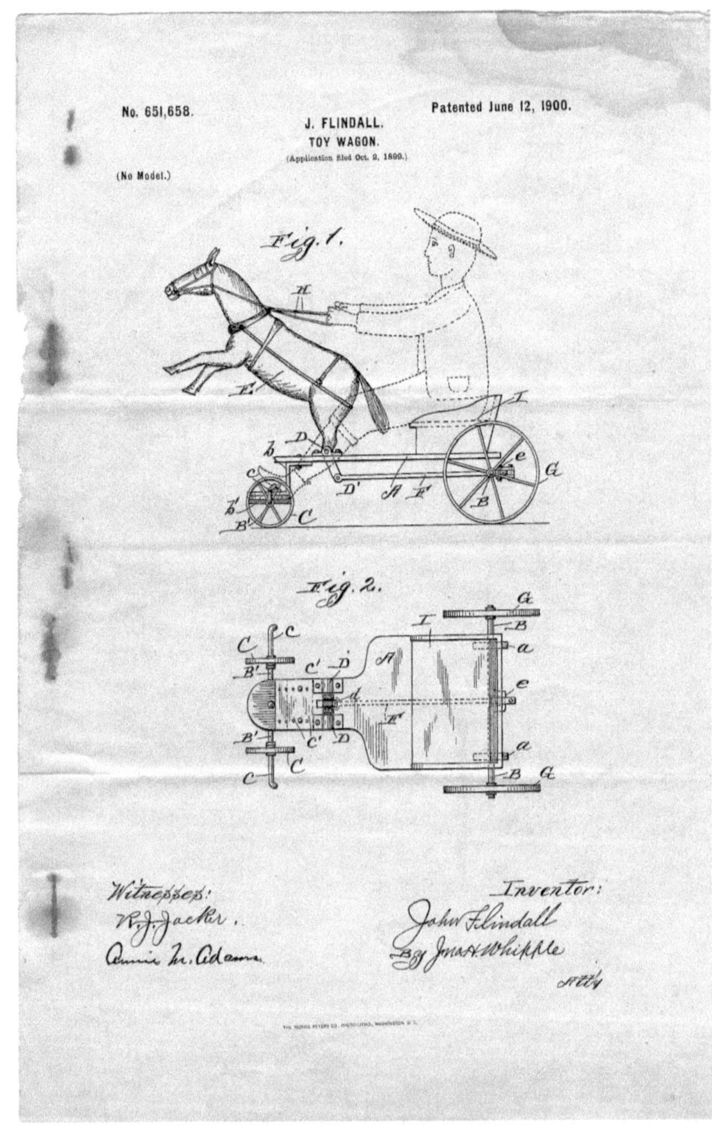
No. 651,658.
Patented June 12, 1900.
J. FLINDALL.
TOY WAGON.
(Application filed Oct. 2, 1899.)
(No Model.)
Fig. 1.
Fig. 2.
Witnesses:
Inventor:
John Flindall
By Jonas H. Whipple
Att'y

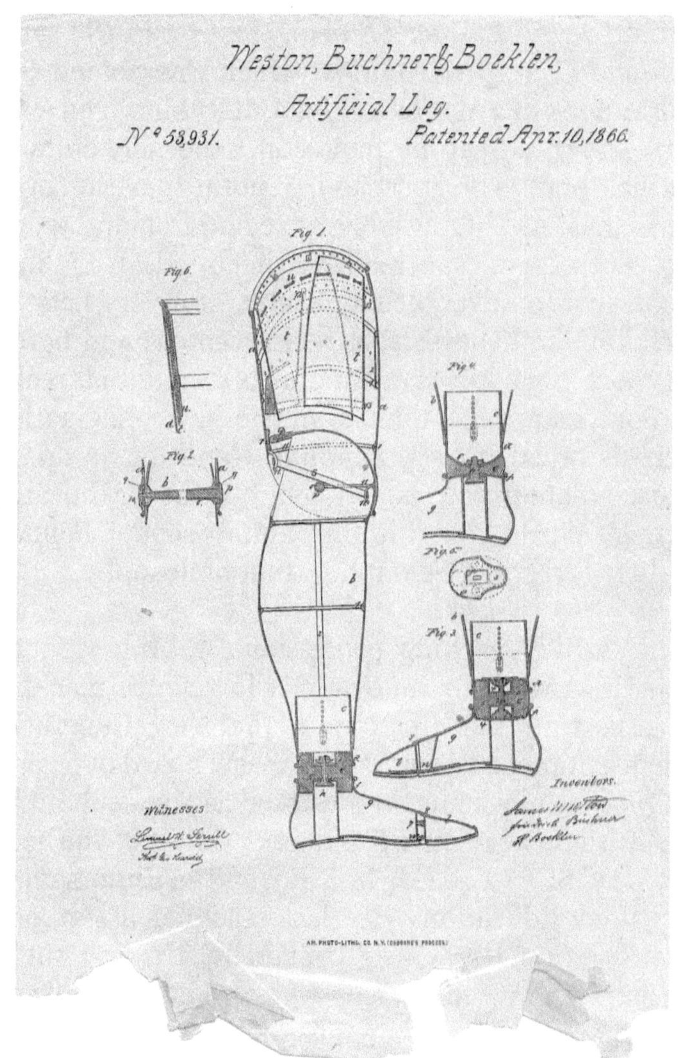

Weston, Buchner & Boeklen,
Artificial Leg.
Nº 53,931.
Patented Apr. 10, 1866.
Fig 1.
Fig 6.
Fig 2.
Fig 4.
Fig 5.
Fig 3.
Witnesses
Inventors.

About the Author

Before writing *Inga's Amazing Ideas*, Ann Rubino wrote *Emmet's Storm*, the first in the series of *Floyd County Chronicles*, which was named Best STEM Book of 2017. As a child Ann Rubino read every book she could get her hands on, especially enjoying the little orange biographies her mother gave her every Christmas. As she raised her five children and taught those of others, she realized that modern children are drawn to adventure, but often of a very fictional sort. Yet she believes that real inventors and heroes are more engaging than fictional ones and that reality is more marvelous than even the best imagination. Her historical novels, *Peppino, Good as Bread*, an intermediate-grade novel about a boy who lives through World War II in Italy and its sequel, *Peppino and the Streets of Gold* follow that philosophy.

While teaching professionally, Ann won the OHAUS Award for innovations in science teaching; took part in the creation of the New Generation Science Standards; sat on the review board of *Science & Children* magazine; and worked as a consultant for the Museum of Science & Industry, Chicago. She holds her MT(ASCP), B.A.Ed., M.S. Ed. and an Endorsement in Gifted Education. Her last teaching assignment was as adjunct at Lewis University, training future teachers. In retirement, she reviewed many children's books for the Recommends division of *Science & Children* and continued her work on the review board.